Water: A Very Short Introduction

'John Finney has written an engaging and compelling account of the puzzling, surprising, yet vital properties of the liquid that comes out of our taps. With one simple geometrical tool as a guide, he takes the reader through successive caverns of ideas, and leaves us peering into the corners that he and his colleagues have yet to explore.'

Tom McLeish, Professor of Physics, Durham University

VERY SHORT INTRODUCTIONS are for anyone wanting a stimulating and accessible way into a new subject. They are written by experts, and have been translated into more than 40 different languages.

The series began in 1995, and now covers a wide variety of topics in every discipline. The VSI library now contains over 400 volumes—a Very Short Introduction to everything from Psychology and Philosophy of Science to American History and Relativity—and continues to grow in every subject area.

Very Short Introductions available now:

ACCOUNTING Christopher Nobes
ADVERTISING Winston Fletcher
AFRICAN AMERICAN RELIGION
   Eddie S. Glaude Jr.
AFRICAN HISTORY John Parker and
   Richard Rathbone
AFRICAN RELIGIONS Jacob K. Olupona
AGNOSTICISM Robin Le Poidevin
ALEXANDER THE GREAT
   Hugh Bowden
AMERICAN HISTORY Paul S. Boyer
AMERICAN IMMIGRATION
   David A. Gerber
AMERICAN LEGAL HISTORY
   G. Edward White
AMERICAN POLITICAL HISTORY
   Donald Critchlow
AMERICAN POLITICAL PARTIES
   AND ELECTIONS L. Sandy Maisel
AMERICAN POLITICS Richard M. Valelly
THE AMERICAN PRESIDENCY
   Charles O. Jones
THE AMERICAN REVOLUTION
   Robert J. Allison
AMERICAN SLAVERY
   Heather Andrea Williams
THE AMERICAN WEST Stephen Aron
AMERICAN WOMEN'S HISTORY
   Susan Ware
ANAESTHESIA Aidan O'Donnell
ANARCHISM Colin Ward
ANCIENT ASSYRIA Karen Radner
ANCIENT EGYPT Ian Shaw
ANCIENT EGYPTIAN ART AND
   ARCHITECTURE Christina Riggs

ANCIENT GREECE Paul Cartledge
THE ANCIENT NEAR EAST
   Amanda H. Podany
ANCIENT PHILOSOPHY Julia Annas
ANCIENT WARFARE Harry Sidebottom
ANGELS David Albert Jones
ANGLICANISM Mark Chapman
THE ANGLO-SAXON AGE John Blair
THE ANIMAL KINGDOM
   Peter Holland
ANIMAL RIGHTS David DeGrazia
THE ANTARCTIC Klaus Dodds
ANTISEMITISM Steven Beller
ANXIETY Daniel Freeman and
   Jason Freeman
THE APOCRYPHAL GOSPELS
   Paul Foster
ARCHAEOLOGY Paul Bahn
ARCHITECTURE Andrew Ballantyne
ARISTOCRACY William Doyle
ARISTOTLE Jonathan Barnes
ART HISTORY Dana Arnold
ART THEORY Cynthia Freeland
ASTROBIOLOGY David C. Catling
ATHEISM Julian Baggini
AUGUSTINE Henry Chadwick
AUSTRALIA Kenneth Morgan
AUTISM Uta Frith
THE AVANT GARDE David Cottington
THE AZTECS David Carrasco
BACTERIA Sebastian G. B. Amyes
BARTHES Jonathan Culler
THE BEATS David Sterritt
BEAUTY Roger Scruton
BESTSELLERS John Sutherland

For more information visit our website

www.oup.com/vsi/

John Finney

WATER

A Very Short Introduction

OXFORD
UNIVERSITY PRESS

OXFORD

UNIVERSITY PRESS

Great Clarendon Street, Oxford, OX2 6DP,
United Kingdom

Oxford University Press is a department of the University of Oxford.
It furthers the University's objective of excellence in research, scholarship,
and education by publishing worldwide. Oxford is a registered trade mark of
Oxford University Press in the UK and in certain other countries

Published in the United States of America by Oxford University Press
198 Madison Avenue, New York, NY 10016, United States of America

British Library Cataloguing in Publication Data
Data available

Library of Congress Control Number: 2015933894

ISBN 978-0-19-870872-8

Printed and bound by
CPI Group (UK) Ltd, Croydon, CR0 4YY

*In memoriam Roger Stone (1911–79) and
Desmond Bernal (1901–71)*

Contents

Preface

Small and apparently insignificant events can have major consequences. In my case, it was a life-changing comment without which I would never have embarked on this *Very Short Introduction*.

In the 1950s, my school class was 'doing' *solids, liquids, and gases* with our physics teacher Roger Stone. Lots of stuff on gases and solids, but we never seemed to get to liquids. 'Please sir. What about liquids?' 'Ah! That's a very interesting question, Finney. Liquids are very difficult. But there's a guy at Birkbeck College in London who's trying to do something about them.' Then the bell rang for the end of the class.

Some years later, once I'd got through my physics degree and was wondering if I should have a go at research, these comments resurfaced. So off went a letter to the 'guy at Birkbeck College'. A fascinating, very informal interview followed, and I ended up working with Desmond Bernal on 'trying to do something about' liquids.

Half a century ago, we really did not understand how molecules are arranged in liquids. We didn't have a clear idea what their structures were—mostly they were thought of as either disordered crystalline solids or dense gases. But now, thanks to advances in experimental techniques such as neutron and X-ray scattering, allied to high-powered computing, we do know a great deal about the structures and properties of not only simple liquids but also of particularly interesting, more complex ones like the water on which our existence depends.

In this book, I attempt to introduce the underlying structural features of water that give its liquid, solid, and gaseous phases the properties they have, and also touch on some of the many areas in which these properties are important in different fields of science. But there are still things we don't fully understand. Water has revealed some of its 'secrets'. But much remains to be done before we can assert that we understand water fully.

Acknowledgements

All scientists are indebted to the many others with whom they have worked or discussed ideas, so it is impossible to acknowledge everyone who has in some way contributed to any piece of work.

However, I am particularly grateful to Austen Angell, Phil Ball, Paul Barnes, Jeremy Baum, John Bouquiere, Daniel Bowron, Graham Bushnell-Wye, Ian Cherry, Roy Daniel, Pablo Debenedetti, John Enderby, Michael Falk, Felix Franks, Alphons Geiger, Julia Goodfellow, Dušan Hadži, Andreas Hallbrucker, Bertil Halle, Kersti Hermansson, Werner Kuhs, Mogens Lehmann, Terry Lilley, Peter Lindley, Colin Lobban, Thomas Loerting, David Londono, Alan Mackay, Dominik Marx, Biljana Minčeva-Šukarova, George Neilson, Ivar Olovsson, Phil Poole, John Quinn, Paolo Radaelli, Valerie Réat, Christoph Salzmann, Hugh Savage, Francesco Sciortino, Neal Skipper, Jeremy Smith, Alan Soper, Robin Speedy, Gene Stanley, Peter Timmins, Jacky Turner, Robert Whitworth and Joe Zaccai without whom there would be major things missing from this book. All the errors are, of course, mine.

List of illustrations

Chapter 1
Water, water everywhere...

Hydrogen, helium, and oxygen are the most abundant elements in the universe. So with helium being essentially chemically unreactive (not for nothing is it called an inert gas), it's hardly surprising that there's an awful lot of H_2O—water—about. We know that there's a lot of it on the Earth as liquid, ice, and vapour; in fact it's the most abundant compound on our Earth's surface. But in the universe as a whole, it's one of the commonest molecules after hydrogen (H_2) itself. It occurs not only as gas in interstellar space and as ice on some other planets and moons in our solar system, but also in the atmospheres of some stars, including our own sun. It is also abundant as a solid ice phase adsorbed on grains of dust, in asteroids and meteorites, and in the cores of the comets that visit us from time to time.

How did water arise from the Big Bang and how did it get to Earth? We're quite confident that we can answer the first question. We're still arguing about the second one. It is thought that less than two minutes after the Big Bang, the temperature was—at about three billion degrees Celsius—low enough for protons (the nucleus of hydrogen) and neutrons to come together to form the nuclei of the lightest elements: helium, lithium, and boron. Once the temperature had cooled to around 4,000°C, the nuclei were able to grab hold of, and retain, electrons; atoms of hydrogen and these other three elements

1

were thus formed. This set the scene for the elemental make-up of today's universe: about three-quarters of it is hydrogen, about a quarter is helium, with everything else constituting merely 1 to 2 per cent of the total.

But where's the oxygen?

We had to wait a very long time for oxygen to turn up. In the gas of matter resulting from the Big Bang, there were regions where the gas density was slightly higher. Gravity then exerted its influence, leading to condensation into 'clumps' of matter that eventually formed stars. Within these starry nuclear furnaces, hydrogen nuclei fused together to form helium. Further fusion processes resulted in heavier elements such as carbon and oxygen—but unfortunately only the *isotope* ^{15}O, which is made up from the 8 protons (which define the element) plus 7 neutrons. ^{15}O is a rare isotope of oxygen—the ones that occur most frequently are ^{16}O, ^{17}O, and ^{18}O (containing 8, 9, and 10 neutrons respectively). Interestingly, it was only in 1957 that a *nucleosynthesis* process was found that could produce these more common oxygen isotopes.

The star would continue to 'burn' until it has used up all its 'nuclear fuel'. Its core would then become unstable, collapsing under its own gravity, and resulting in a spectacular explosion—a supernova. The energy of that event would be such that new nucleosynthesis processes are triggered that result in heavy elements such as uranium and thorium. All the debris, containing pretty much everything in the periodic table, would then be scattered into the interstellar void.

If we look out into space, we find the results of these supernovae explosions and what time has done to them subsequently, for example in the spectacular molecular clouds we can observe with a good telescope. In addition to still-dominant hydrogen, we find other gases such as carbon monoxide, methanol, and ammonia. And water.

There are also very small grains of solid material such as silicates and carbon (even diamond), often with a coating of ice. As the material within these molecular clouds begins to clump, it flattens out into a rotating disc. Most of the material in the disc gathers together at the centre of the disc to form a star, while that which is further out forms the store from which the planets are formed. This planet formation is thought to occur by accretion of gas and dust grains to form small rocky bodies (*planetesimals*) that collide with each other, and cohere under the force of gravity to form larger bodies. Most of the remaining material is used up in these collisions, with the smaller bodies being 'eaten up' by the larger ones, resulting in the planets, but not yet as we know them. Perhaps more like molten rock, but already containing much of the material that we now have on Earth.

How did the water get here?

There are several mechanisms by which the water we have on Earth might have got here. One popular theory is that it came through impacts of 'wet' bodies such as comets, which contain large amounts of frozen water. Furthermore, we have already noted that the dust grains that accreted to form the Earth had icy coverings that came from the gas bathing the dust grains. It is possible that much of that water was retained on accretion, and as the mass of water in the region of the 'proto-Earth' was much greater than the amount of water that was eventually accreted, this could be a significant mechanism by which water came to us.

The capture of gas from the solar nebula—the cloud of gas and dust from which the sun and the rest of the bodies in the solar system formed about 4.5 billion years ago—is another mechanism that has been proposed, though it is thought unlikely that much water can have arrived this way. In addition to water from cometary impacts, collisions with asteroids could also have added water. This theory is difficult to test, but the evidence we do have suggests this is unlikely to be the main source.

As to what the actual mechanism was, the jury is still out; a comfortable position to take is that the three main mechanisms listed here ('wet accretion', gas from the solar nebula, and late impact of comets and asteroids) all contributed to the water we have on Earth. But how much came from each is unclear and a subject of current research. For example, the Rosetta Mission took samples of the surface of a comet in 2014 partly to look for evidence to help to resolve this long-controversial issue.

What water did next

The distribution and history of Earth's water from the time of its arrival to the present day is one of the most intractable problems in geochemistry. We can, however, do some informed speculation. During the later stages of the formation of the Earth, the colliding planetesimals would have deposited enough energy to at least partly melt the Earth to produce a magma ocean (ejected material from which may have subsequently formed our moon).

Together with water, other volatile molecules such as hydrogen (of course), nitrogen, and oxides of carbon, would also have been incorporated into the magma ocean during the accretion process. Then, as the magma cooled and began to solidify, the volatiles would have been ejected to form a primitive atmosphere consisting mostly of hydrogen, carbon dioxide, and water vapour. The gravitational field of the Earth would be insufficient to keep hold of the light hydrogen, which was gradually lost. As the temperature of the Earth cooled low enough for the water vapour to condense into clouds, the rains came—and stayed—so that the oceans were formed.

Although uncertainties remain, a plausible geochemical case can be made for modern patterns of crust formation, erosion, and sediment formation having been established within about a hundred million years of the Earth's formation, though whether plate tectonics had begun by then is an open question. By then, we may well have had a primitive atmosphere and large bodies of water—the liquid water oceans.

...nor much of a drop to drink

Fast forward to the present day when, for pretty obvious reasons, we know a lot more about the water on our planet. Though certainly not everything—for example we don't know with any degree of certainty just how much we have. We know that some 70 per cent of the Earth's surface is ocean, while around 5 per cent of it is ice-covered as frozen ocean, Antarctic and Arctic ice caps, and glaciers. We have a pretty good idea of how much water we have in the oceans and rivers, as ice, in groundwater, rocks, and organic matter: about 1.9 billion billion tonnes (or 1.9 billion cubic kilometres), of which some 70 per cent is oceanic salt water.

Surprising as it may seem, there's a great deal of water also in the Earth's mantle—the rocky silicate layer between the Earth's crust and its outer core. Most estimates of water content in the mantle range from 0.2 to 2.5 billion billion tonnes—the higher figure being equivalent to around two Earth oceans' worth. Though there are even higher estimates than that.

This means there's an awful lot of it about. However, most of it is unavailable to us as the fresh water we need: forgetting the water in the mantle, only around 2 per cent of these 1.9 billion billion tonnes is fresh water. And as nearly two-thirds of this is locked up in the ice caps, taking that out of the equation means less than 1 per cent of the fresh water on Earth is potentially available to us. Considering 90 per cent of this is in groundwater which we would be silly to completely mine, the amount of fresh water we can reasonably access is around 0.1 per cent of the grand total. This may still be an apparently large amount of water (1.9 million billion tonnes), but the small fraction does focus the mind on the need to wisely look after our easily accessible fresh water sources.

Why do we need it?

The short answer: for maintaining our climate and keeping us—and other living things—alive.

Both these statements relate to complex processes, in which water in all its three phases is critical. We need an adequate amount and quality of water, and for this the hydrological cycle (evaporation and transpiration followed by precipitation) is critical as a water purification mechanism—though as about 75 per cent of the 'recycled' water falls as rain on the oceans, it's not available to us without (energetically and financially) costly desalination. And although the actual amount of water in the atmosphere at any one time is small, it plays a central role in mediating our weather and climate through a complex set of feedback loops.

Important too is the polar ice cover, which reflects back a significant amount of the incident radiation from the sun. Without this ice cover, more heat would be absorbed by the Earth. In addition to warming the land masses, this would also result in the warming of the oceans, with consequent effects on ocean currents such as the Gulf Stream which is a major player in maintaining the temperate zones of northern Europe.

The environmental factors that mediate our climate are complex and not yet fully understood—we are talking about very subtle balancing acts involving a number of different processes. But in many of these, water in all its three phases is at the heart. And as we shall see later, its particular properties in the liquid (for example the relatively large amount of heat it takes to raise the temperature of a given amount of water by one degree Celsius) and as solid ice (which has a lower density than water) are very pertinent to its influence on our climate.

Psychological as well as physiological needs

On a more personal note which I hope will ring bells more widely, there are reasons why I and many others need water which are additional to its physical functions of keeping me alive, and maintaining the environment in a state that allows my continued physical existence.

In order to keep sane, I have to disappear every now and then into the hills and the valleys, and on to the coastal mudflats and isolated islands to soak up the atmosphere of what's out there in the wilder places that we have left. I need this, not only to refresh myself spiritually, but also to enable me to clear my mind to let me think through difficult problems, scientific or otherwise. And water has been—and continues to be—a major force in shaping those inspiring landscapes.

For example, the deep U-shaped valleys that host dramatic waterfalls falling into them from hanging valleys were carved by ice. *Liquid* water is also a major shaping force of the landscape: rivers and streams have slowly ground away material to produce V-shaped valleys, in the process eroding away every year billions of tonnes of sediment that can build up coastlines. And those coastlines themselves can be continually worn away by the water in the oceans.

With its ability to acidify through dissolving carbon dioxide, water has been central to the formation of the features we experience in limestone areas: from the beautifully varied landscapes of dry valleys, to sink holes that swallow up the streams of surface water that hollow out dramatic underground caverns such as in the Carlsbad National Park in New Mexico. These subterranean treasuries host stalactites and stalagmites, and other superb water-formed natural sculptures that attract cavers into the depths. While those of us who find it too claustrophobic to go underground in confined spaces can enjoy the—again water-carved—limestone pavements characteristic of, for example, the Irish Burren, or the dramatic tower-like limestone hills in south-eastern China that have inspired classical Chinese painters.

Water, then, is important for our souls as well as for keeping us alive. How it does the latter we will look at in Chapter 6.

Water in geology

In looking at the influence of water on the Earth, I have only just scratched the surface. Not only is water a major influence in forming the surface topography of the planet, but it is also critical in what happens in planetary interiors, as well as in the location and eruptive style of volcanoes. I have already commented that the amount of water in the Earth's mantle might be enough to refill the Earth's ocean several times over. So what is it doing down there and how might its presence such a long way under our feet affect us?

Let's take just one example. When an oceanic plate is pushed beneath a continental plate (a process known as *subduction*), it drags down water with it (for example within hydrous minerals). As it sinks, the descending plate is heated by processes such as friction between the plate and the mantle and by the effect of pressure. If this heating is sufficient, the plate will begin to melt, and as the molten material will be lighter than the crystalline rock, it will tend to float upwards.

Water can play a crucial role here in that, when under pressure, the melting point of rocks is significantly reduced by the presence of water. For example, dry granite 20 km below the surface melts at 1,000°C, while 'wet' granite turns molten at only 650°C. This (lighter) fluid magma can then rise and, on reaching the surface, may result in explosive volcanic activity.

This may seem odd at first, but it's well established that those dramatic volcanic eruptions depend on water down below. Thus many of the geological features that are the result of volcanic activity have water in their formation history. And not just on Earth: some parts of the Moon's mantle are thought to contain as much water as does the Earth's upper mantle.

But what exactly *is* water?

I have blithely assumed that water is a molecule made up of one oxygen atom bound to two hydrogen atoms. This conclusion is based on a number of major advances in science including atomic theory itself and the recognition of the existence of elements. Getting to this stage of knowledge about water took a long time: it wasn't until the second quarter of the 19th century that water was recognized as H_2O.

How might water have looked to our forebears much further back in history? With water being all about them, and the only compound on Earth that exists naturally in all three phases—gas, liquid, and solid—it's hardly surprising that it has grabbed the attention, and been the inspiration of, philosophers, poets, musicians, and visual artists, as well much more recently physical, chemical, Earth, and biological scientists.

Its ubiquity and importance for sustaining life may have inspired the ancients in their ascribing to water a position of primacy in their philosophy. In the account of the creation in Genesis, the waters were there before the land: 'Let the waters under the heaven be gathered together unto one place, and let the dry land appear'. The Babylonians gave water a critical role in creation, with all lands being sea until the god Marduk intervened to create dry land. In the Qur'an, the prophet Mohammed claims a revelation from God that 'we have made of water everything living'. The Romans believed that spa and mineral waters were put on Earth by God to cure human ills, while in the Christian church, water was seen to bring spiritual benefits at baptism, and holy water could cleanse us of our sins.

The elements of the ancient world

It is generally to Thales of Miletus—one of the seven sages of antiquity who is sometimes called the father of philosophy—that we tend to ascribe the first declaration (around 700 BCE) that water was the basic constituent of everything. Why not indeed when it can be vapourized to form a kind of *air* and frozen to form a kind of *earth*? Surely this could therefore serve as the basis for *all* (solid, liquid, and gaseous) matter?

Others following Thales preferred to give primacy to other 'elements' such as air and fire, with Empedocles in the 5th century BCE perhaps being the first to bring together the four 'classical' elements of the ancient world: earth, air, fire, and water. It may also be to Empedocles that we accredit the first ideas of atoms, with his concept of 'seeds' having four 'complexions' corresponding to the four elements.

Other civilizations took different sets of elements for their bases. For example, the Chinese fixed on *five* elements of earth, wood, metal, fire, and water—perhaps we might note here the importance to the Taoist of the number five. Not surprisingly, water appears as an element in common in these different traditions.

The Alexandrians, probably influenced by contacts with Egypt, Asia Minor, Mesopotamia, and perhaps even India, began to insert practical aspects into their natural philosophy (including the realization that steam could do work, though without building a real steam engine). Despite these moves towards pragmatism, the Aristotelian ideas of the elements (as well as his other ideas on matters as diverse as physics and chemistry to ethics and politics) seemed to freeze thought processes, so preventing Europe from

breaking out of this intellectual straightjacket for around two
millennia.

Leonardo da Vinci did begin to use experimentation as a tool for
the development of understanding of many things. For example,
by concluding that water evaporated into the clouds, which
gave rise to rain, he got part of the way towards an understanding
of the hydrological cycle. But the idea of water as elemental
continued to survive. However, with the development of
experimentation in the 17th century suggesting possible
transformations between various pairs of the four elements, it
became increasingly difficult to fit observations into that classical
four-element model.

Water in the beginnings of modern science

In the latter half of the 17th century—the period that saw the
beginnings of modern chemistry—even Robert Boyle, who had the
insight to formulate a fundamental law concerning the behaviour
of gases (Boyle's Law), was unable to break away from the earlier
'elemental' ideas.

Though he accepted that water could be transformed into organic
matter (a mid-17th-century experiment of Jan Baptista van
Helmont showing the uptake of water by a growing tree having
been interpreted as the transformation of water to wood), he still
regarded water as an element. In his experiments on combustion,
however, the seeds of unravelling the composition of water are to
be found. Without identifying the gases as such (that was to come
later), he obtained hydrogen gas ('impure air') by dissolving iron
filings in acid, and burned it in air. He was thus perhaps the first
to unite hydrogen and oxygen to form water—but unknowingly.

The final resolution came some hundred years later following the
discovery by Joseph Priestley of a gas that promoted combustion
that was later to be named oxygen. But bound intellectually by the

existing conceptual framework, he also was unable to see the light. In this same period, Henry Cavendish, a rich 'gentleman scientist' who weighed the Earth, produced 'inflammable air' from the action of acids on iron, zinc, and tin, and noted that this gas burned easily in air. But as with Priestley, in interpreting the results of these experiments, he also was unable to break out of the conceptual framework of the time. (This didn't prevent his successors naming a prestigious physics laboratory at Cambridge after him.)

So by 1775, both components of water had been isolated. But they still had not been established as elements rather than 'impure airs'. The 1770s saw three experiments that united hydrogen and oxygen. These were performed by John Warltire (a colleague of Priestley), Pierre Joseph Macquer (a colleague of Antoine Lavoisier in France who we will meet again shortly), and James Watt, who was later to do some rather interesting work with the steam engine but was then a laboratory technician of Joseph Black, the Scottish chemist who discovered carbon dioxide. Priestley also performed this experiment in 1781.

All produced a residue of water, but that was not seen as particularly remarkable as the condensation of water from the air is commonplace. Cavendish got even closer to the 'answer' when he observed in 1781 that one of the 'impure airs' (oxygen) united with 'inflammable air' (hydrogen) in the quantitative ratio of 1:2. H_2O indeed, but not recognized as such. A very near miss.

This whole story is long and convoluted, and illustrates the problem of trying to interpret new results in terms of old concepts. It also shows a reticence of some workers to acknowledge the work of others, a characteristic that unfortunately can still be found in present-day research, be it scientific or otherwise.

To cut this long story short, it took Lavoisier in France to come to the correct conclusion and bring order to the situation by realizing that water was a compound of hydrogen and oxygen. This he was able to do only by jettisoning the conceptual framework—the intellectual straightjacket—that the British scientists had been unable to break out of. In doing so, he named the two gases concerned: *hydrogène* for its ability to generate water, and *oxygène* for its acid-forming ability—though of course we now know that not all acids contain oxygen. Indeed, he further verified his conclusion by 'splitting' water and recovering the resulting hydrogen.

The confirmation of the actual formula was to have to wait another fifty years, during which Dalton developed the beginnings of modern atomic theory. He initially got it wrong, making an assumption that compounds should consist of one atom of each of the constituents, from which he assigned to oxygen an atomic weight of 7 (though it would have been 8 had he had more accurate data). It took the Swedish chemist Jöns Jakob Berzelius to make the final jump in 1826. By correcting oxygen's atomic weight to 16, water finally became H_2O.

So now we know what this substance is.

Chapter 2
The water molecule and its interactions

The key to the way water behaves in all its different guises—vapour, liquid, crystal, and (yes) glass—lies in the molecule itself. So once we understand the nature of the molecule, we can then think about how it interacts with other water molecules to explore its behaviour and properties as a gaseous, liquid, or solid material. We can study its molecular sociology. Similarly, if we can understand how a water molecule interacts with other kinds of molecules, we can explore the properties and behaviour of the wide range of chemical, physical, and biological systems in which water is involved.

Water in solitary

So what does a water molecule look like on its own?

The answer really depends on what particular property you might want to focus on. For example, its structure is often thought of in terms of a 'ball and stick boomerang' arrangement of its constituent oxygen and hydrogen nuclei (see Figure 1, ignoring for now the arrows and plus and minus signs). The balls represent the nuclei of the atoms; the sticks represent the chemical bonds between them. In these *covalent* bonds, some of the electrons (the *valence* electrons) from each atom are considered to be shared between the different nuclei making up the molecule.

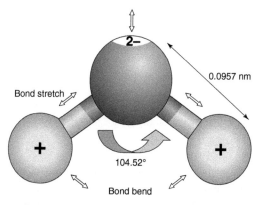

1. Ball and stick representation of a water molecule. Arrows show how the atoms vibrate; + and − signs show where the electric charge is concentrated.

This depiction emphasizes the geometry of the molecule, which is not a bad thing to do as this geometry has a major influence on the way water molecules interact with each other and with other molecules. The central oxygen nucleus is separated from each of its bonded hydrogen nuclei by a distance of about a tenth of a billionth of a metre: 0.1 nanometres (nm).

Noting that many other three-atom molecules, for example carbon dioxide (CO_2), are linear, at first it's a little surprising that the water molecule is 'bent'. There are good quantum mechanical reasons for this, which when we do the calculation properly explain why the HOH angle is about 104.5°. What is particularly interesting about this angle is that it is close to not only the tetrahedral angle of around 109.5°, but also to the internal angle of the pentagon (108°). Remember these angles. They will come in handy later when we begin to rationalize the structures of both ice and liquid water.

But this static picture, though it will help us to understand the geometry of the way water molecules interact, is too simple, for no molecule is ever static. It vibrates. The simplest way of thinking about its vibrational motion is to imagine each O–H bond as a spring, with the hydrogen and oxygen atoms at each end of the spring able to oscillate—as indicated schematically by the arrows along the O–H bonds of Figure 1. In addition to this *bond stretching* mode of vibration, the hydrogens can also oscillate towards and away from each other so as to vary the HOH bond angle. This *bond-bending* mode is also illustrated by the relevant arrows in Figure 1.

As in all molecules, as the temperature is increased, the vibrational motions increase. And *vice versa*: as we reduce temperature, the vibrational motions decrease. However, even at the absolute zero of temperature (0 Kelvin, or –273°C), some vibrational motion remains.

This is a quantum mechanical effect called *zero-point motion*. For systems containing light atoms, these motions can be considerable and potentially significant in determining some behaviours. Thus, as the water molecule contains two of the lightest atoms—the hydrogens—even at absolute zero, the molecule retains significant vibrational energy. It may seem a little odd to emphasize this point when considering the behaviour of water under normal ambient temperature conditions, but as we shall see in Chapters 5 and 6, these zero-point motions may be important in some processes that are chemically and biologically relevant.

Charged-up water: enter the electrons

This description so far tells us about the shape, size, and internal motions of water molecules, but doesn't tell us anything about how they attract each other, or attract (or repel) other kinds of molecules. To get some insight into this, we need to consider the effects of the electrons in the molecule. Some of these bind together the oxygen and hydrogen nuclei to form the water molecule. But to understand how the molecules are likely to interact with each other, we need to look at the water molecule as a distribution of electric charge.

The molecule itself is electrically neutral. Each hydrogen contains 1 proton, each oxygen has 8. So an H_2O molecule contains 10 positively charged protons, a charge which is balanced by 10 (negatively charged) electrons. If we do the quantum mechanical calculations (or make the relevant experimental measurements), we find that the charge is not uniformly distributed within the molecule. Rather, there is a degree of separation of the positive and negative charges.

The simplest representation of this charge separation is to assign to each of the hydrogen atoms a small positive charge (the plus signs on Figure 1), with a balancing negative charge 'on' the oxygen that is twice the magnitude of the charge on each hydrogen. Rather than placing it at the centre of the oxygen, quantum mechanical calculations suggest it's better to think of the negatively charged region as a single 'lobe' of charge on the 'top' side of the oxygen. The charges can be thought of as centred on the corners of an isosceles triangle: the molecule thus has an overall approximately trigonal charge geometry.

A molecule in which positively and negatively charged areas are separated will be describable quantitatively in terms of a set of *electrical moments*. These are useful to us here in that, once we know their values, we can work out how the molecule responds in an electric field, in particular an electric field created by the electrical charges on neighbouring molecules. In other words, the electrical moments enable us to calculate how the molecules might attract—or repel—each other.

The most important of these moments is the *dipole moment*. Its magnitude for water is 1.85 Debye, the unit being named after the famous Dutch physicist Peter Debye. A highly appropriate naming, as Debye's first major scientific contribution was the application of the dipole moment concept to the charge distribution of asymmetric molecules. This dipole moment value is quite high, but is not out of line for a small molecule.

If we place the molecule in an electric field, be it an externally applied one or that deriving from the electrical charges on neighbouring molecules, it will respond through the interaction of its dipole moment with that external electric field (a point which is relevant to water's very useful ability to act as a good solvent which we will raise again in Chapters 5 and 6). However, the external electric field will have an additional effect on the water molecule: it will further distort the charge distribution round the molecule in such a way as to increase the dipole moment.

We therefore say that the molecule has a significant *dipole polarizability*, with its dipole moment being increased by the presence of an external field. Again, this behaviour is important in understanding water's behaviour as a good solvent. However, we should note that this is a property common to many small molecules. Moreover, neither the magnitude of water's dipole moment nor its dipole polarizability, though both quite high, are out of the ordinary for a small molecule.

Before leaving our discussion of the electrical properties of the water molecule, we should note that molecules will repel each other when they get so close to each other that their electron distributions begin to overlap. The simple bent dumb-bell picture of the molecule in Figure 1, though it gives us a picture of where the charge is concentrated, tells us nothing about the spatial extent of the electron distribution.

A better picture is given in Figure 2. Based on high-quality quantum mechanical calculations, this shows us that the molecular shape can be described reasonably well by a slightly non-spherical surface. Experimental data from high-resolution crystal structure analyses support such a slight non-sphericity. Though this aspect

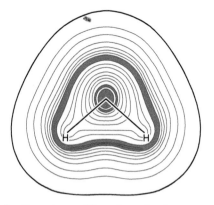

2. Electron density contours of the water molecule.

of the water molecule is not often emphasized, it can help to explain in a relatively simple way the quite complex water structures we often observe, for example in biological systems.

So what's special about the molecule?

We have seen that the water molecule is made up of two hydrogen atoms chemically connected (covalently bonded) to one oxygen atom. The molecule is not linear, but bent, with the HOH angle of its average geometry being, at around 104.5°, close to both the tetrahedral angle and the internal angle of a planar pentagon. It is continually vibrating internally. As it contains two light atoms (the hydrogens), its zero-point motion is significant. The water molecule's overall charge distribution can be thought of as being near-trigonal. This charge distribution gives the molecule a significant dipole moment, and the ease with which this distribution can be distorted results in a significant dipole polarizability, though neither the dipole moment nor the polarizability is out of line with the values found for many other small molecules. Finally, the molecular shape as defined by its electron density distribution is slightly, though significantly, non-spherical.

In summary, looked at by itself, it is apparently an unremarkable small molecule.

Getting together

Water molecules interact with each other through a type of interaction called *hydrogen bonding*. This is a much weaker interaction than the covalent bond that links the hydrogen and oxygen atoms in the water molecule (and atoms in molecules generally). It is however stronger than the *van der Waals* interaction that weakly attracts atoms or molecules which do not have a significant dipole moment, for example molecules such as methane CH_4.

The actual nature of the hydrogen-bonding interaction remains a little controversial. Though early ideas tended to think of the interaction as purely an *electrostatic* one, being determined by the charge distribution on the molecules, later theoretical work has often suggested there is a small degree of electron transfer between interacting water molecules, hence implying some covalent character. Whether we should consider the hydrogen-bond interaction in the case of water as partly covalent remains controversial. But to understand the structure and properties of water in general, we don't need to worry about resolving this controversy.

What is not controversial is the strength of the interaction between two water molecules. It is intermediate between that of a van der Waals interaction and a covalent one. The order of magnitude of the interaction (numerically about 20 Joules per mole) can be put into some sort of context by noting that this figure is equivalent to about ten times a typical thermal fluctuation at room temperature.

The relative magnitudes of intermolecular attractions and thermal fluctuations are relevant to the cohesion of an assembly of molecules at a given temperature. In the case of water, this order of magnitude difference between the energy of a hydrogen bond and of a typical thermal fluctuation means water remains liquid at ambient temperature. The thermal fluctuations are not strong enough to break apart enough hydrogen bonds to vapourize the liquid. In contrast, the strengths of the interactions between similar small molecules such as hydrogen sulfide (H_2S) are much weaker; they are therefore more easily broken by thermal fluctuations and hence such liquids will boil at considerably lower temperatures than does water (-60°C in the case of H_2S).

This relative strength of the hydrogen bond with respect to the perturbing thermal fluctuations gives a simple rationalization of one of the so-called anomalies of water: that it is liquid at ambient temperature, while many other molecules of similar or greater molecular mass such as H_2S remain gaseous.

But there is more to the water–water hydrogen bond than just its strength. Of particular interest—and importance—are its directionality and the number of molecules that like to interact with each other.

It would seem very reasonable for the positive hydrogen of one molecule to be attracted to the negatively charged region of the oxygen on another molecule. We would therefore not be surprised to find a hydrogen pointing towards the back side of the oxygen of a neighbouring water molecule. However, having a negative charge that is twice the positive charge on a hydrogen, the oxygen won't be satisfied with hydrogen bonding to only one other molecule. It still has some 'pulling power' left to attract a hydrogen on yet another water molecule. In fact, there is an understandable preference for each oxygen to attract the hydrogens from two neighbouring water molecules.

Furthermore, this 'central' water molecule also has its own two hydrogens to satisfy, so these will tend to interact with the negative region of each of two further water molecules.

The end result is shown in Figure 3, in which each water molecule is surrounded ideally by four hydrogen-bonded neighbours. In two of these interactions, the central water molecule will act as hydrogen-bond *donor* through pointing its (positively charged) hydrogens at a negative region of each of two neighbouring waters. In the other two interactions, the central molecule will act as a hydrogen-bond *acceptor* of two neighbouring waters that point their hydrogens towards the negative region of the central water molecule. Quite a pleasing set of symmetrical interactions that keeps all the molecules happy. On this simple argument, we can build up the four-coordinated motif of Figure 3. The four neighbours form an ideally *tetrahedral* arrangement around the central molecule.

This ideal local four-coordinated arrangement is found experimentally. Crystallographic studies on structures containing other molecules together with water (hydrates) show that water

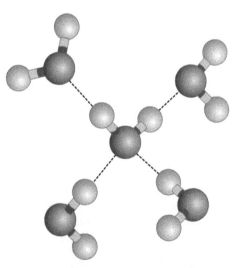

3. A four-coordinated water molecule showing the ideal tetrahedral arrangement of the first-neighbour environment of a water molecule.

hydrogens almost always participate as hydrogen-bond donors where negatively charged regions of neighbouring molecules are available and accessible. So, even though the negative region is only a single lobe of negative charge, the availability of space to accommodate two donor hydrogens in the doubly negatively charged region of the water oxygen will push the molecules towards the tetrahedral coordination seen in this four-fold motif. The hydrogen-bond donor–acceptor ratio is 2:2. High-quality theoretical calculations also reproduce this local structure.

Look carefully at this arrangement of five water molecules and commit it to memory! It is the key to understanding water—its structure in its various guises, its properties, and its behaviour in all sorts of circumstances and environments. So we will refer to it frequently (after a while you may begin to feel ad nauseam) as we use it to explore the ways in which water molecules interact in *condensed phases*—starting with the familiar crystal form you find in your freezer or on the ski slope.

Chapter 3
Water as ice(s)

No matter where you find it occurring naturally on Earth—in your freezer, falling as snow in your garden, or in an arctic glacier—ice is a *crystal*. At the molecular level, this means that the molecules that make up the crystal are arranged in a regular repeating pattern. Put slightly differently, we can imagine a small number of molecules arranged inside a box that we call the *unit cell*. To form the crystal, we just replicate that unit cell by placing other molecule-containing unit cells next to it in three dimensions to form the extended crystal.

In performing this repetition, we have to have a unit cell that, when repeated in three dimensions, fills all the space without leaving any gaps. This restricts the number of shapes of boxes that can form valid unit cells—clearly a cube can pack together with other identical cubes to fill three-dimensional space, but a tetrahedron, for example, cannot (try it!).

Moreover, when a unit cell containing molecules is repeated, the interactions between the molecules across the unit cell boundaries must be consistent with the way the molecules interact with each other. Thought of in terms of a jigsaw puzzle, not only must the shapes of the pieces fit with each other, but so also must the picture elements match across the joins.

Your normal ice

Moving now to consider the molecular structure of ice itself, we can ask a simple question. How can we connect together a number of water molecules that interact in the manner indicated in the four-coordinated motif of Figure 3 to form a regular repeating structure that fills the correct volume (which we know from measuring the density of ice) at a given temperature—say just below the melting point of ice of 0°C?

To simplify the picture a little, let's forget the hydrogen atoms for the moment and represent the structure in terms of the oxygen atoms of each water molecule. The answer to our question is then given in Figure 4, where the structure of 'normal' ice is shown from two different views.

This structure has a number of interesting features. First, each water molecule is indeed four-coordinated, consistent with what we learned in Chapter 2 about the way the molecules prefer to interact with one another. Second, the water molecules form (puckered) six-membered ring structures, with the OOO angle formed by three neighbouring water molecules being essentially the tetrahedral angle of 109.5°. Being very close to the average water molecule HOH angle of 104.5°, this means that the water molecules are very 'comfortable' in this arrangement: if I showed the hydrogen atoms, we would see that they are only very slightly off the direct line between neighbouring oxygen centres. Put slightly differently, the hydrogen bonds between neighbouring molecules are only slightly bent. Third, all the (average) hydrogen-bond lengths in the structure are nearly identical.

Finally, as is obvious in one of the orientations shown in Figure 4, the structure is very open—we can see through open channels in the structure. Even though the amount of 'empty space' in the structure is exaggerated by representing each water

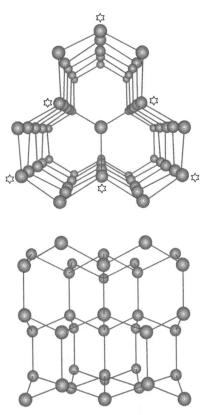

4. The structure of 'normal' ice looked at from two perpendicular directions. For clarity, only the oxygen centres of the water molecules are shown.

molecule by its oxygen centre, if I replaced each oxygen by the space-filling electron density representation of Figure 2, there would still be almost enough volume to squeeze in further water molecules into these channels. That this doesn't happen in ice is a consequence of the angular constraints forced on the arrangement by the directionality of the water–water interaction. Even if we did try to squeeze further molecules into these spaces, they would not be able to bond with the molecules already there.

Were we to force the matter and stuff other water molecules into these holes, we would completely fill the space (forming a *close-packed* arrangement) and approximately double the density of the crystal. The density of the real ice structure is therefore only approximately half the density it would be if the directional interaction requirement wasn't there. As we shall see in Chapter 5 when we come to consider the so-called anomalies of water, this open structure is very relevant to why ice is less dense than—and hence floats on—water.

If this kind of open crystalline structure really is a consequence of the underlying tetrahedral geometry of the four-fold motif of Figure 3, we would expect to find similar crystal structures in other systems where the constituent atoms or molecules interact in a similar tetrahedral fashion. And we do. Examples include the silicates in many rocks, and diamond.

Putting the pressure on

But the ice story doesn't stop here. In fact, the ice we have just discussed I should have really called ice Ih. The (Roman) numeral 'I' because it was the first kind of ice we knew, and the letter 'h' because of its hexagonal symmetry (as clearly seen at the molecular level in Figure 4, and also in the hexagonal symmetry of individual snowflakes when you catch them on a cold surface). This labelling implies there must be other kinds of ice. Indeed there are.

For example, if I cool down ice Ih to about −30°C and apply about 3,000 atmospheres pressure, I find a different *phase* of ice, ice III. This is still a crystal, in which the molecules still interact through our familiar four-fold motif (still remember it?). If I now cool this ice III to about −40°C I get yet another phase, ice II. Initially surprisingly perhaps, if I apply pressure to *liquid* water at room temperature, at just over 10,000 atmospheres I get yet another phase, ice VI, and yet another phase (ice VII) on increasing the pressure further to above around 13,000 atmospheres.

In fact, at the latest count, there are seventeen (some would say eighteen) confirmed crystalline phases, which means ice crystals with seventeen (or eighteen) different structures. That is, seventeen or eighteen different ways in which the water molecules are arranged to form different crystal structures.

All these structures are based on the four-fold motif of Figure 3, and the temperature and pressure conditions that are needed to make most of these phases are shown in the *phase diagram* of water of Figure 5. For example, if I take a point at 20°C and about 1,000 atmospheres pressure, I find I am in the region of stability of liquid

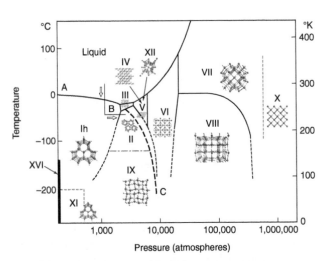

5. The phase diagram of ice, many of showing the phases we expect to see at different temperatures and pressures. The thumbnail images show the different structures. The dashed line BC is an extrapolation of the melting line AB to higher pressures.

water. If I stay at the same pressure and lower the temperature to about −40°C, I cross a line (a *phase boundary*) into ice Ih. If I now increase the pressure further, to, say about 3,000 atmospheres, I will cross another phase boundary into ice II. And so on.

The structures of many of these phases is shown in the thumbnail images on the phase diagram. It's interesting to look at some of these in a bit more detail to see the kind of mischief water molecules can get up to.

Molecular versatility

First, let's consider again what happens when I increase pressure on ice Ih to about 3,000 atmospheres at about −30°C to form ice III. Increasing the pressure will compress the ice so that it occupies less volume. This will increasingly strain the molecular arrangement of ice Ih (Figure 4) until it reaches a point when it is

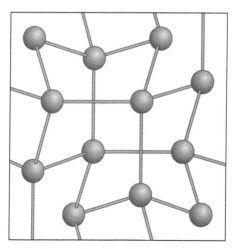

6. The structure of ice III, showing only the oxygen centres of the water molecules.

more comfortable for the molecules to rearrange themselves to form a different structure that occupies less volume.

How the molecules have rearranged to occupy less volume (about 20 per cent in this case) is hinted at in the view of ice III shown in Figure 6: we can pick out at least one *five-fold* ring in the unit cell shown (how many can you find?). And the molecules in this ring will occupy on average less volume than they would in the six-fold rings of ice Ih.

Rearrangement into shorter rings is one way in which water molecules play games to occupy less volume when put under pressure. The four-fold motif is maintained, though in order to accommodate smaller rings the hydrogen bonds will be increasingly strained and distorted. In the case of ice III, the average OOO bond angles vary between 88° and 148°. There is also increased variation in bond lengths. Whereas in ice Ih these were all nearly equal at 0.276 nm, in ice III they range between 0.271 nm and 0.283 nm. So to be able to occupy less volume while still retaining the overall tetrahedral coordination motif, the hydrogen bonds have had to distort to facilitate the formation of smaller ring structures that occupy less volume.

But this isn't the only way the molecules respond to the need to occupy less volume as pressure is raised. As we increase the pressure even more, there's a limit to how much we can strain the hydrogen bonds to form shorter rings. So to get round this problem, the molecules cooperate to do something remarkable: they make use of some of the empty space we have already commented on in an ingenious way by 'threading' hydrogen bonds through six-fold rings.

An example is ice IV (Figure 7). Again, the four-coordinated motif is fully retained, and there is a further increase in the range of bond length distortion to 0.274 nm to 0.324 nm: the longest bonds are, not surprisingly, those related to the 'threaded' bonds which are 'stretched' by the surrounding intermolecular

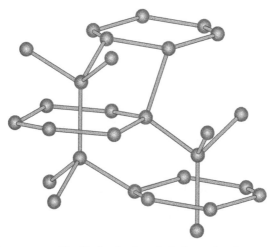

7. The structure of ice IV, showing how the hydrogen bonds can 'thread' through rings.

repulsions. The OOO bond angle variation is only two-thirds that of ice III, suggesting the threading relieves some of the angular stress of the bonding.

This idea of threading is taken to extremes in the higher-pressure ices VI, VII, and VIII. Each of these three structures can be considered as two identical water molecule networks that fully interpenetrate to fill the available volume. This is most easily illustrated by ice VIII, for which the way the structures interpenetrate is shown in Figure 8. But the four-coordinated motif is still retained in each of the interpenetrating sub-lattices, and as the individual networks are good tetrahedral ones, the driving force to bond distortion has almost vanished. Where there is strain is from the small expansion of each network from its ideal to accommodate the volume of the water molecule in the 'holes'—otherwise there isn't quite enough space to fit in the 'other' network. This results in a stretching of the hydrogen bonds by around 5 per cent to 0.288 nm.

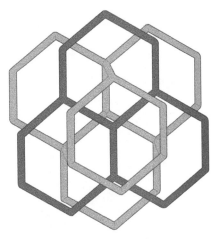

8. The structure of ice VIII. To emphasize the interpenetrating networks, only the hydrogen-bonded links are shown.

We might note in passing that the structure of each of the interpenetrating networks is the same as that of diamond. It is interesting to speculate whether two diamond networks could be persuaded to interpenetrate in a similar way. And if so, would the resulting material be even stronger? Or prettier?

Summing up

From this brief examination of four ice structures, we can see that the ordered arrangements of water molecules all show four-coordinated geometry, consistent with the four-coordinated motif of Figure 3. There are however variations in hydrogen-bond lengths and OOO angles induced by pressure when forming the high-pressure ice structures.

The bond-bending ability of the water molecule seems to be relatively 'easy', accommodating apparently quite comfortably significant deviations from the ideal near-tetrahedral value.

When the volume available to them is reduced by increasing the pressure, the water molecules exploit the possibilities of hydrogen-bond distortion by arranging themselves in an impressively large number of different ways to fill space. But they retain their preferred four-coordinated local arrangement in all these structures.

When considering the molecule itself, we concluded half way through Chapter 2 that it looked a rather unremarkable one. After observing the molecule's versatility in structure forming in the various ices, perhaps we should begin to revise that opinion. The four-coordinated motif seems to be remarkably resilient, maintaining its integrity over a very wide range of temperatures and pressures. Perhaps this versatility has relevance to its properties and function? This is an issue we will return to in considering both the so-called anomalies of water (Chapter 5) and its much-vaunted biological role (Chapter 6).

'Ghostly' ices and 'missing' hydrogens

You may have noticed some apparent inconsistencies in what's been said so far about the ices and how I have illustrated the different structures. I will now try to clear these up. In doing so, we will throw light on aspects of both crystalline and liquid water that turn out to be not only interesting, but also quite important in understanding their properties and behaviour.

First, in the phase diagram of Figure 5, the pressures and temperatures at which two particular phases are shown to occur (ices IV and XII) are indicated not by an area on the figure but by arrows pointing into the region of stability of ice V.

This is because these two phases are *metastable*: in the pressure and temperature region indicated, ice V is the stable phase and ices IV and XII can only be prepared by very specific procedures, some of which may involve adding small amounts of impurity which appear to stabilize the metastable structures. In passing, to perhaps illustrate the attractions (or frustrations) of working on ices, I might mention that although ice XII was first observed as a metastable phase in the ice V region, it has since been found to occur as the stable phase at lower temperatures.

Ices XVI and XVII, discovered respectively in 2014 and 2016, are also metastable at normal pressure. Interestingly, XVI is postulated to be stable under *negative* pressure, i.e. under tension. So its region of stability may be to the left of Figure 5. The structures of these recently discovered phases are also particularly interesting. We will meet them again in a different guise in Chapter 6.

Second, in illustrating the structure of ices Ih, III, and IV in Figures 4, 6, and 7, I showed only the oxygen centres to emphasize the ways the molecules are linked together. I used an even simpler representation showing only the hydrogen-bonded links between water molecules when emphasizing the interpenetrating sub-lattice structure of ice VIII (Figure 8). Why didn't I show the hydrogens?

Furthermore I have been a bit non-committal in saying that there are seventeen or eighteen known ice structures, while only thirteen are shown on Figure 5. Where are the others and why aren't they shown on that diagram? The answers to these questions are related.

Disorderly hydrogens and the ice rules

The simple answer to the second question is that I didn't show the hydrogens in ices Ih, III, and IV because we don't know where they are. This may sound a strange admission, but to explain it I reproduce the tetrahedral motif of Figure 3 in sextuplicate in the left half of Figure 9. Why repeat it six times? Look carefully at each of the six images and it will be apparent that they are not the

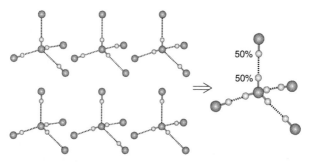

9. The six possible arrangements of hydrogens in the four-fold motif (left-hand side), and the average structure that would be seen by a diffraction experiment (right-hand side).

same; they vary in the positions of the hydrogen atoms—or, what is in essence the same thing, the orientation of the central molecule.

Each of these figures shows a valid four-coordinated local structure. Focusing on the hydrogen atoms, we can convince ourselves that their arrangements are consistent with a set of rules relating to ice structures first set down in a classic 1933 paper by J. D. Bernal and R. H. Fowler which was so fundamental to our understanding of water that I will be referring several times to it. These rules—which really express geometrically the known chemistry—state that, in ice:

1. Each oxygen atom is covalently linked to two hydrogen atoms. This is essentially a definition of the water molecule.
2. The two hydrogen atoms in each water molecule each form a hydrogen bond with one other water oxygen, so that there is precisely one hydrogen between each pair of oxygen atoms.

A consequence of the application of these two rules is that each water molecule links to four neighbours through hydrogen bonding, as we have seen in the four-coordinated motif that has been central to our discussion so far.

Now looking at our six examples in the left-hand side of Figure 9, we see that the central water molecule is in a different orientation in each example. And consequently the orientations of the neighbouring four molecules need to be different to be consistent with the Bernal–Fowler rules.

When we now think of how we might arrange many water molecules into a crystal, this hydrogen requirement makes things just a little more complicated. Taking any of the crystal structures we have discussed in terms of the positions of the oxygen centres of the water molecules, we are faced with the possibility of two different kinds of structure when we add the hydrogens.

In the first of these, the water molecule *orientations* as well as their positions are *ordered*. This means that in each unit cell, equivalent water molecules are in the same orientation. In the second kind of structure, they are *disordered*: in each unit cell, equivalent water molecules are *not* in the same orientation. This difference is illustrated for simplicity in a two-dimensional analogue in Figure 10, from which we can see that the ice rules are observed in both the ordered and disordered structures.

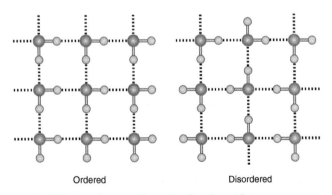

Ordered   Disordered

10. The difference (in a two-dimensional analogue) between a hydrogen-ordered (left) and hydrogen-disordered (right) structure.

While formally we should perhaps talk about this order/disorder being *orientational*, in that it relates to the relative orientations of the water molecules, we often speak of it as *hydrogen* order/ disorder, and tend to use the two different terminologies interchangeably.

So why is this apparently obscure piece of detail relevant to the questions posed earlier?

First, some of the ices shown in the phase diagram of Figure 5 are orientationally disordered, some are ordered. When they are orientationally disordered, then as the structure we measure averages over all the equivalent molecules in the crystal, we don't know at which end of a hydrogen bond we should expect to see a hydrogen atom: in half the cases it will be attached to the central water molecule in our four-coordinated motif; in the other half it will be attached to the neighbouring molecule.

Our crystallographic structure determination (best done by neutron scattering) will show *all possible* positions for the hydrogens, so our structure will appear to have hydrogens at *both* ends of the hydrogen bond. But as each hydrogen bond in reality can only accommodate one hydrogen, each of these hydrogen sites will be occupied with only a 50 per cent probability (see the right-hand part of Figure 9). So in a disordered structure, we gain nothing from showing the hydrogens—hence my dropping them in the earlier figures.

Of the ices shown in our phase diagram, ices Ih, III, IV, V, VI, VII, XII, XVI and XVII are orientationally disordered, while ices II, VIII, IX, and XI are orientationally ordered. Ice X is something rather different that we will mention later. Ice IX has—thankfully— nothing whatsoever to do with Kurt Vonnegut's 'Ice Nine' phase in his 'Cat's Cradle' novel, a phase that had apocalyptic consequences, though we shall get some echoes of similar apocalyptic warnings when we look at the polywater episode in Chapter 7.

Ices behaving badly

We are now getting a little closer to finding out why there are apparently four (or five) phases missing from the phase diagram of Figure 5. And in discovering the reason, we will get to grips with some further interesting issues related to important aspects of the behaviour of both ice and liquid water.

Consider first the disordered phases. There is a well-known principle of physics that order increases as temperature is reduced. Surely, then, as we reduce temperature, the water molecule orientations should change until an ordered ice phase is produced?

Indeed this does happen easily in the cases of disordered ices III and VII: as we reduce their temperatures, the water orientations change to form the ordered phases IX and VIII respectively. But the other disordered ices are much less obliging. For example, ice XI, the ordered version of the familiar ice Ih, is *not* formed simply by cooling. Similarly, no matter how slowly we cool them, the other disordered phases IV, V, VI, and XII all behave badly and stay disordered. They are stuck in their disordered structures even at very low temperatures.

This does not mean we are breaking the third law of thermodynamics that tells us everything should be ordered at absolute zero. It merely means that we are unable to access the ordered structures by simple cooling. There is something preventing the ordering process taking place.

And that something is due in part to the Bernal–Fowler rules.

Bringing ices to order

Consider again the two-dimensional analogue of a small part of a disordered ice crystal in Figure 10. The arrangement clearly obeys the Bernal–Fowler rules: each water molecule consists of an oxygen atom covalently linked to two hydrogens which are each hydrogen-bonded to neighbouring waters, and each water oxygen is also hydrogen-bonded to two other neighbours through the hydrogens of those other waters.

If we are to order this array of waters, we will have to rotate a number of molecules to take up different orientations. For example, a rotation of 180° of the central molecule will get it into the orientation shown in the ordered structure. But it is prevented from making that rotation by the Bernal–Fowler restrictions: there are already hydrogens from neighbouring molecules on the linkages that this rotation would need to place its own hydrogens on, so it cannot make that move. To make further progress we need to find a mechanism whereby these sorts of rotations are possible.

One suggestion might be to say that a number of *coordinated* moves have to be made simultaneously by a number of molecules, in such a way that the Bernal–Fowler rules remain fulfilled. In principle this is possible, but the energy required to activate such a cooperative process is large, and so unlikely to be available at low temperatures. Another is to examine the possibilities of defects in the ice structure that may facilitate the required orientational movements.

That such defects must exist, we know from the electrical properties of ice: it conducts electricity through the motion of protons so there must be some mechanism whereby the hydrogens can move in the structure when an electric field is applied, for example from left to right in Figure 10. Once such defects are present—even in small numbers—orientational motions will become possible that may be able to propagate through the crystal.

Two types of defect are now generally recognized in ice structures: *orientational* defects (or Bjerrum defects named after their

11. The different kinds of defects in ice structures and the possible movement of an H₃O⁺ defect.

proposer), and *ionic* defects. As illustrated in the top half of Figure 11, there are two kinds of Bjerrum defect: a *D-defect* (from the German *doppelbesetze*, meaning doubly occupied) and an *L-defect* (from *leere*, meaning empty). These descriptions should be self-explanatory from the figure. The ionic defects are simply H_3O^+ and OH^-, formed by the transfer of a proton (hydrogen) from one molecule to a neighbour. This ionization is a familiar process in liquid water, with both ions occurring at a concentration of 1 in 10 million water molecules at neutrality.

Either kind of defect can 'free up' rotational motions, as indicated in the bottom half of Figure 11 for the H_3O^+ ionic defect, with migrating point defects leaving behind a trail of water molecules whose orientations have been changed. We thus now have a possible mechanism to reorient molecules so that the disordered

40

ice crystal can reach its preferred ordered low-temperature structure.

These *intrinsic* point defects do indeed seem to be sufficient for ices III and VII to order on cooling to respectively ices IX and VIII. However, the other disordered ices still resist ordering for some reason that we don't yet understand. To persuade these to order, it has been found necessary to introduce *extrinsic* point defects by adding very small amounts of an acid or a base; when coupled with careful, slow cooling, this seems to promote the ordering process.

In the case of the ice Ih, doping with small concentrations of potassium hydroxide (one KOH in about 500 waters) releases an ordering transition to ice IX at about −200°C. In contrast, doping with acid rather than alkali (HCl at a much lower doping level of about 1 in 50,000) seems to be required for ices V, VI, and XII to order. Careful cooling of each of these three phases then results in their recently discovered ordered phases: respectively ices XIII, XV, and XIV. These are three of the phases 'missing' from our phase diagram (Figure 5).

Ice X and beyond

The highest-pressure ice we have considered so far is ice VII (or its orientationally ordered form ice VIII). As the water molecules in this double interpenetrating lattice structure are already as closely packed as they can be, what happens if I continue to apply pressure to this structure?

First, I will squeeze the molecules closer together. One result will be that the distance between the hydrogen of one molecule will get closer to the oxygen of its hydrogen-bonded neighbour. If I continue to squeeze the crystal further, there will come a point at which each hydrogen atom will be at the same distance from both its 'owner' water molecule and its 'hydrogen-bonded' neighbour. At that point, the hydrogen won't be able to distinguish between its 'parent' and 'neighbour' molecule. The hydrogen bond will have become symmetric, and the Bernal–Fowler rules requiring there be distinct H_2O molecules will be irrelevant. I will have in effect a crystal of hydrogens and oxygens rather than of distinct water molecules.

The pressure at which such a transition to what has been termed ice X occurs has been an active area of both theoretical and experimental investigation for many years. Though spectroscopic features that might indicate a transition to a symmetric ice have been observed at pressures of around half a million atmospheres, there has to date been no conclusive identification of ice X.

With even greater pressures, what might happen to our poor water molecules? Ultimately the material should eventually become metallic. In fact, experiments on ice at up to 6,500K and over 4 million atmospheres pressure have revealed two superionic (high proton mobility) phases. These may be relevant to the interiors of the ice giant planets, but they don't really impinge on our understanding of the behaviour of our earthly water. So we will leave this discussion here and return to Earth.

The curious case of cubic ice

So far, I have—perhaps infuriatingly—refused to say whether there are seventeen or eighteen known ice phases. In now explaining why I have been so indecisive, we will come across some other interesting issues relevant not only to ice on Earth, but to ice elsewhere in the solar system.

Let's return to the upper picture of the structure of ice Ih in Figure 4. We can envisage this structure as built up by connecting together layers of puckered hexagonal rings on top of each other. The way I have stacked these layers for ice Ih is such that, in the view shown in the top half of the figure, the rings lie on top of each other—hence the open channels I have previously remarked on.

Furthermore, if we look at two neighbouring stacked layers, we see that these are mirror images of each other; this is perhaps most easily seen in the lower view in Figure 4. If we label each layer so as to recognize this difference, we might call these two layer types A and C.

This ACACAC arrangement we find in ice Ih is, however, not the only way to stack these hexagonal layers. There is another way.

Consider the top two layers in the upper picture of Figure 4. Imagine disconnecting the hydrogen bonds joining the top two layers (under the starred oxygens) and rotating the top layer by 60°. The labelled molecules can no longer be reconnected to the layer below. However, we can get the top layer into a position where it can reconnect to the layer below by moving it downwards to a position in which the starred oxygens can reconnect. In doing so, we will have brought some molecules over the open channels in the original stacking. Let's label this reoriented top layer type B. This is most easily seen by using models, so don't worry if it's not clear—just note that there are three possible ways of labelling each layer that are consistent with the way puckered hexagonal layers of water molecules can hydrogen-bond to each other: A, B, and C.

We have already noted that ice Ih results from the ACACAC stacking and that this has hexagonal symmetry. In contrast, the ABCABC stacking gives us a different structure. As water with this stacking would have cubic symmetry, it has been labelled ice Ic, or more affectionately cubic ice. This is the 'missing' ice structure that we have so far been cagey about. Why have we been cagey about it? Quite simply because attempts to make a pure sample of it have failed for over half a century.

Until the mid-1980s, ice Ic was indeed thought to be a real phase of ice, though it didn't seem to occur in a unique area in the phase diagram. The material that was called cubic ice was made in a number of ways, usually by warming up some of the high-pressure ices at ambient pressure, and was sometimes thought to be another metastable phase. However, there were features in the X-ray and neutron diffraction data that are used to determine crystal structures that didn't quite look right.

To cut a very long story short, what had previously been called cubic ice has been found experimentally to be a structure in which the stacking of the hexagonal layers is irregular. Not ABCABCABC etc., but a disordered arrangement of layers, for example ABCBCACBACAB…and so on. Hence ice Isd—stacking disordered.

Such stackings are all fully hydrogen-bonded, built up of our basic four-coordinated motif. But they are not 'good' crystals in the strict sense. They are perhaps better described as our old friend ice Ih but with faults introduced in stacking the layers on top of each other during crystal growth. Such stacking faulted structures are known elsewhere in materials which have structures related to that of ice, for example silicon carbide. So although this kind of solid structure is interesting, it is not unique to water.

So if it isn't a 'good' crystal, why should we be bothered about it, especially when its structure was rather messy to explain? Quite simply, because it is relevant to the way in which water crystallizes to ice, and the form in which water is found elsewhere, both in the Earth's atmosphere and possibly elsewhere in the solar system.

Let's take the crystallization point first. When a liquid crystallizes to a crystalline solid, there is an energy barrier to the formation of the first small nucleus of the crystal. It turns out that the energy barrier to form small nuclei of 'cubic ice' is lower than that required to form the stable hexagonal ice Ih. However, as that nucleus grows, the stacking faulted structure becomes unstable with respect to ice Ih and so transforms to the regular hexagonal ice Ih structure. And if we grow ice in confined spaces, for example in microporous glass, we again find that the first few layers can be examples of the cubic arrangement. That the stacking faulted structure is indeed a metastable one is demonstrated by heating the structure to a temperature at which the defects we discussed earlier that allow orientational motions are thermally activated. The stacking faults are annealed out and we are left with our familiar ice Ih.

Moving away from the surface of Earth, there is evidence of the presence and possible importance of 'cubic ice'. For example, there are indications that snow crystals start their lives as 'cubic' nuclei. Furthermore, the ice in the high-altitude clouds formed by crystallization from water vapour may sometimes contain examples of the 'cubic' structure. Christoph Scheiner first described in 1629 the eponymous halo that is sometimes seen at 28° from the sun or moon when either shines through one of these clouds. This 28° angle is consistent with this hypothesis. Moving completely away from Earth, there is also evidence for cubic ice being a constituent of cometary ice, and also for being a significant astrophysical player elsewhere in the universe.

In my five years teaching condensed matter physics, I used the ice phase diagram (figure 5) to illustrate certain points. Each year, as new phases were discovered, I had to change the phase diagram I used! True to form for ice, a similar thing has happened since this little book was first published: not only has ice XVII been discovered and two superionic structures found under very high pressures, but real cubic ice has finally been prepared pure. A versatile system indeed, with perhaps more phases to be found.

Chapter 4
Water as a liquid—and as glass(es)

Can a liquid have a structure?

A characteristic property of a liquid is that it *flows*. In contrast, a crystalline solid doesn't flow (except on very long time scales which we can ignore for the present). We know that in a crystal, each molecule is centred on a particular fixed position, and so we have no problem in conceiving of a crystal having a structure. In a liquid, however, for the liquid to flow, the molecules must move. So can we talk about a liquid having a structure?

The answer is, of course, yes we can. Otherwise we wouldn't be trying to understand how water structure explains many of the properties of the liquid. But in developing the idea of what we mean by liquid structure, we can usefully look again at how we thought about the structures of ices.

For simplicity, I implicitly assumed in Chapter 3 that the water molecules in the ices were static, in order to emphasize the ways the molecules connected together through hydrogen bonding to form the different structures. Yet in Chapter 2, we noted the bond bending and stretching vibrations that occur in the isolated molecules; these motions will continue—though at frequencies slightly perturbed by the intermolecular interactions with neighbouring molecules—in the ices.

Moreover, the molecules themselves will oscillate about their mean positions.

But as the structure is simply defined by the average positions of the hydrogen and oxygen atoms making up each water molecule, and the motions will be around those equilibrium positions, these motions will not affect the basic crystal structure. If we took a snapshot of the structure with an exposure time longer than the typical vibrational times, we would see a slight blurring around the mean positions of the atoms, but the average structure defined by those mean positions would not be affected. Similarly, the kind of rotational motions of water molecules that we discussed in Chapter 3 will not affect the average structure.

So the ice structures we have discussed are *vibrationally averaged* structures, or *V-structures*. They are the average positions of the molecules that we would see if we could take a molecular-level snapshot of the crystal with an exposure time longer than the characteristic time scale of the molecules' vibrational motions.

We can consider the structure of a liquid in a similar way. First, let's take a snapshot with an exposure time that's longer than that of the individual molecular vibrations, but shorter than that which would see individual molecules move from their 'instantaneous' positions. This would tell us where the molecules are before they had chance to diffuse to new positions. Now a few seconds later, let's take another snapshot with the same exposure time. This will show us another 'instantaneous' arrangement of molecules that in detail is different from the first one, as the molecules will have moved between the snapshots.

However, if we look carefully using the right descriptive tools, we can observe structural features that are common to all the snapshots we take—for example the average number of molecules at a given distance from each molecule. If we now average these structural properties, we can characterize the V-structure of the liquid.

Simple liquids first!

Before we look at the structure of water, we need to step back and think about the problem of liquid structure more generally. In doing so, we are following in the footsteps of one of the great polymaths of the 20th century, J. D. Bernal, who was not only in the vanguard of the development of crystallography generally, but was also instrumental in its use in trying to understand the structures of biological molecules such as proteins. In fact it was his interest in how biological systems operate at the molecular level—the study of which is central to molecular biology—that led him to think about the structure of water. Realizing that water was ubiquitous in living systems, he argued that if we want to understand biological functioning, we need to understand water.

In fact, as we noted in Chapter 3, Bernal, together with R. H. Fowler, produced in 1933 the first serious paper on the structure of water. The story has it that this was written following an overnight discussion between the authors at a fog-bound Moscow airport. Considering the influence this paper has had—and continues to have—on our understanding of water and aqueous solutions, that airport delay was perhaps one of the most productive transport disruptions in the history of science!

Despite the seminal nature of that paper, Bernal was later to find his approach then to the structure of liquid water unsatisfactory. It was, he said, 'frankly one of crystal structure, trying to picture water structure as that of a mixture of the analogous four-coordinated structures of quartz and tridymite'. That model was too ordered: 'This was ultimately to prove rather a delusive approach, postulating a greater degree of order…in the liquid than actually exists there'.

As he himself admitted, the structure of water was a difficult problem. This was particularly so because at the time there was little understanding of the structures of much simpler liquids. So he argued that first he needed to understand simpler liquids: 'It is not worth tackling complicated liquids until we understand simple ones'. And he went on to develop an approach to simple liquid structure that enabled him over two decades later to build up a picture of water structure that agreed with the experimental data of the time. As we shall see, those ideas are also very relevant to what we have learned about water's structure from today's state-of-the-art experimental structural investigations of the liquid.

Liquids are liquids—not gases or crystals

The structure of liquids is a problem that has proved very difficult to approach theoretically. This is in stark contrast to our understanding of the structures of crystals and gases. So it's understandable that gases and crystals have tended to be used as departure points for trying to understand liquid structure.

As we saw in Chapter 3, in an ideal crystal the positions of molecules in a unit cell can be specified, and the structure of the extended crystal reproduced simply by repeating that unit cell in three dimensions. The structure of a real crystal can be approached through looking at the effects of perturbations from this ideal, for example thermal motions and the presence of defects. The underlying periodicity of the structure dramatically simplifies the calculation of the physical properties of a crystal from known molecular parameters.

An ideal gas is in the opposite corner from an ideal crystal in that the instantaneous positions of non-interacting point atoms are random. Just as we approach a real crystal by perturbing the regularity of the ideal crystal, we approach the real gas by looking at the perturbations resulting from giving the point atoms a size and allowing them to interact at a distance. As a gas is compressed, the molecules are pushed closer together, and consequently interact more strongly. So long as the density remains reasonably low, we can predict quite well the structure and properties of a simple gas.

It is therefore not surprising that early attempts to understand liquid structure approached the problem from both of these extremes: by considering the liquid to be either a highly disordered crystal or a high-density gas. Although each approach can go some way to predicting some properties of liquids, neither can satisfactorily explain the experimental data.

Put simply, treating liquids as disordered solids assigns them *too much* order. If we treat them as dense gases, we can't assign them *enough* order: as the gas becomes denser and approaches condensation to a liquid, the theory cannot handle the complexity of the interactions that occur in the condensed liquid.

Bernal found these 'crystal-like' and 'gas-like' approaches to liquid structure fundamentally unsatisfying. For a simple liquid—one of atoms that could be regarded as attractive soft spheres—he wanted a simple model. Perhaps because he was a crystallographer and hence looked through a structural eye, he wanted a more concrete picture of the structure of the liquid: 'some kind of theory of liquids that would be homologous to that of the crystalline solid as well as being radically different in kind, and have a general quality of homogeneity'. Unlike a crystal, he argued, a liquid doesn't contain lines or planes of molecules. So a satisfactory model must not assume such structural entities that are just not present in the liquid.

Bernal's solution was a very visual and inherently simple idea. And an idea that subsequent work has demonstrated to be essentially correct as a reference model for real liquids.

Heaps and piles

You will recognize the four or five bottom layers of Figure 12 as a 'crystal' of steel balls. The hard, spherical 'atoms' are arranged in a

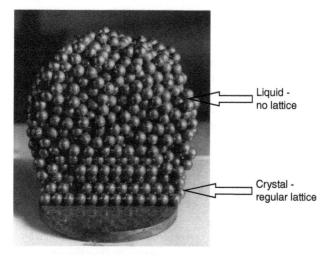

Liquid -
no lattice

Crystal -
regular lattice

12. Bernal's model of the arrangement of spheres in a simple liquid (above) compared with the ordered regular arrangement in a crystal (bottom).

regular, repeating pattern, and lines and planes of atoms characteristic of a crystal are apparent. This is a regular *pile* of 'atoms', with the atoms interacting with each other in a way that is consistent with their sphericity and hardness.

In contrast, the upper part of Figure 12 is an irregular *heap* of the same kind of hard, spherical 'atoms'. There is no obvious regularity—there are no planes or long straight lines of 'atoms'. Yet the 'atoms' are still interacting with each other as hard spheres in the same way as they are in the crystal. This disorderly heap is in fact a snapshot of the structure of a liquid of hard, spherical 'atoms'—here of a (time-frozen) liquid of steel balls. As Bernal put it simply and succinctly, the structural difference between a *crystal* and a *liquid* is the same as that between an ordered *pile* and a disordered *heap* of atoms. In both cases, the interactions between the atoms are the same— here hard sphere repulsions. But as Figure 12 shows, the structure of the liquid and the crystal are fundamentally different.

Real simple liquids are, of course, not hard spheres. They are better thought of as soft spheres that attract one another—the interaction between them is called their *intermolecular potential*. And in the case of simple liquids such as argon, it is the repulsive part of the potential that drives the liquid structure. As Bernal was fond of saying, 'the key word in the structure of liquids is the one which Humpty Dumpty used in *Alice through the Looking Glass*: "impenetrability". The actual structure must be determined largely by the form of the repulsive forces between close molecules.'

This model—christened random close packing of equal spheres— may be simple, but later work has shown its essential validity, not only as a model of an ideal simple liquid, but also as a valid starting point for liquid state theory generally. As John Ziman, one of the 20th century's great liquid state theorists has said: 'This simple idea…is now seen to be the key to any qualitative or quantitative understanding of the physics of liquids.'

And those liquids include water.

From packings to networks

This may all be very well as a picture of a liquid of spherical atoms, but how can it help us to understand the structure of water, where the interaction between the molecules—the basic four-fold hydrogen-bonded motif you've committed to memory (Figure 3)—is far more complex than the isotropic spherical interaction between simple atoms?

The extension to water is conceptually straightforward. In Chapter 3 we approached the ice structures by imagining how we could connect together water molecules to occupy a given volume in a manner that was consistent with both the water–water interaction (the four-fold hydrogen-bonded motif) and the existence of a crystal lattice. For an ideal model of the structure of liquid water, we can follow the same conceptual approach, but this time *discarding the constraint of a crystal lattice*. All we need to do is to connect together water molecules so that each connects to four neighbours, taking care to not grow an ordered crystal and to produce a model of the right density. This we can do with little difficulty.

Figure 13 shows a laboratory model that is fully consistent with these conditions: the black balls representing the water oxygen atoms connect to four neighbouring molecules through the white hydrogen spheres. The constraints of the four-fold motif (a consequence of the water molecule's intermolecular potential) are fulfilled, but there is no regular repeating crystal-type organization.

13. A random network arrangement of water molecules. The large black spheres are the oxygen centres, the small white spheres represent the hydrogens.

In analogy to the simple liquid model being termed a *random packing*, this structural model of water is a *random network*. To illustrate the concept further, Figure 14 shows a two-dimensional three-coordinated analogue of both the liquid and crystal structures. Interestingly, the liquid looks like it could (as indeed it does) have a higher density than the crystal, a point we will return to in Chapter 5.

One way of quantifying the difference between the ordered crystal and the disordered liquid is, as we did with the ices, to look at the ring structures. In the two-dimensional analogue of Figure 14, the crystal is made up of only six-fold rings, while the random network has a mix of five-, six-, and seven-fold rings. In three dimensions, as we learned in Chapter 3, ice Ih consists of six-fold rings, while in the ideal liquid water model, a range of rings from four- to seven-membered and above is found.

How does the structure of *real* liquid water match up to this ideal random network model? We now have far better experimental data on the structure of water than Bernal ever had access to.

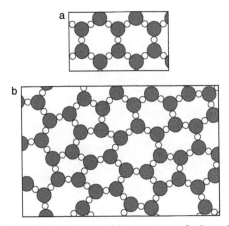

14. Two-dimensional analogues of the structures of a. ice and b. liquid water. To accommodate the two-dimensionality, the coordination has been reduced from four to three.

Consequently we do now know what liquid water looks like at the molecular level. In summary, just as the random sphere packing model contains the essence of the structure of a simple liquid, so the random network model contains the essence of the structure of liquid water under normal conditions of temperature and pressure.

In my view, the most important structural data on water have been obtained from the scattering of neutrons and X-rays. These methods have a great advantage in that there is a direct relationship between the measured scattering and the actual structure. In contrast, in order to interpret the results from many other experimental techniques, we need to assume some sort of model that relates the measurement to the structure. The interpretation of the data is therefore inherently dependent on the validity of that assumed model. Consequently, if that model is incorrect, then the conclusions we draw from the data will also be incorrect. It is therefore safer to rely on the results of good scattering measurements, which we can now ally with state-of-the-art computer calculations to generate actual structures that are consistent with the experimental scattering data.

The results of many years of such experiments back up Bernal's proposal that the random network is indeed an excellent ideal reference model of the structure of water. We can in fact examine the coordination of each of the water molecules in these sample structures (though setting a criterion for coordination is not straightforward). The results show that, as expected, the four-fold tetrahedral motif dominates. They also tell us how the real structure differs from the reference random network both under normal conditions and as we change temperature and pressure.

Defects in the random network structure

A blow-up of a typical limited region in liquid water, derived from neutron scattering measurements mentioned earlier, is shown in

Figure 15. Although the four-coordinated motif is found to be the dominant coordination in the 60,000 molecule liquid water assembly from which this instantaneous snapshot is taken, a close inspection of the figure shows a good example (labelled A) of three-fold coordination in which there is only one hydrogen of a neighbouring water pointing towards the negatively charged region of the molecule A.

This should not really surprise us. In Chapter 2, we considered the negatively charged region of the water molecule as a single 'lobe' of charge. As the value of that charge is twice that on each hydrogen, a water molecule would prefer to link its negatively

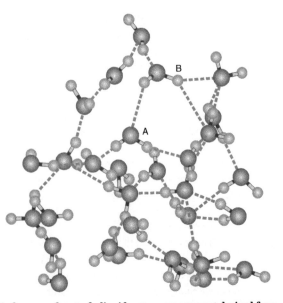

15. A close-up of part of a liquid water arrangement derived from experimental measurements, showing likely hydrogen bonds between neighbours. The orientation is chosen to show examples of a three-coordinated molecule A, and a bifurcated interaction B in which one hydrogen appears to hydrogen-bond to two neighbouring waters.

charged region with two hydrogens. However, when two are not available, it has to make do with just one. In the context of the overall preferred four-fold coordinated motif, the resulting three-fold trigonal arrangement might be thought of as a defect in the ideal random network model.

A further 'defect' can also be seen in Figure 15. Labelled 'B' in the figure is the so-called *bifurcated* interaction in which a hydrogen of one water molecular seems to hydrogen bond to the negative regions of two, rather than the normal one, neighbouring waters. Although in the example shown there is only one hydrogen donating to the bifurcating molecule's negatively charged region, in those cases where that negatively charged region is involved in two 'normal' hydrogen bonds and the molecule's other hydrogen makes a normal hydrogen bond to another neighbour, such a molecule would have a local five-fold coordination.

As we shall see later, these 'defects' from the ideal random network turn out to be quite important in the contexts of both the dynamics of the liquid (see later in this chapter) and its biological relevance (Chapter 6).

When discussing the ices in Chapter 3, we noted that in order to fit into a smaller volume as we increased pressure, the bond angles and bond lengths became increasingly distorted from their 'ideal' values. Similarly, when we look at the OOO bond angles formed by three neighbouring water molecules in the ideal random network model, we also find significant variation from the ideal tetrahedral angle.

If we examine the same bond angle distribution in water structures obtained from experiment, we also see a range of values, though as we would expect for a liquid, the variation is broader than in the ices. Also when discussing ices, we saw how the various structures could be considered in terms of the ring structures that the constituent water molecules formed; as bond angles and bond lengths varied in the high-pressure structures, a range of ring sizes (actually from 4 to 10) became possible.

Similarly, in the water structures obtained experimentally, we find four-, five-, six-, seven-fold and more, ring structures, as there are in the ideal random network model. We might recall here the comment we made in Chapter 2 about the water HOH angle of 104.5° being close to the internal angle of a planar pentagon (108°). So water molecules are likely to feel very comfortable when sitting in five-fold ring structures.

Spatial distributions

Though the four-fold motif is the basis of the structures of both the ices and liquid water, the fact that there are distortions from the ideal in both kinds of structure means that it's useful to have a pictorial way of quantifying the average local structure around a central water molecule. This *spatial density function* is shown in Figure 16 for water at room temperature. The lobes above and below the central water molecule represent population densities of water molecules in those positions and orientations—think of the lobes as being contours representing a particular density of neighbouring water molecules.

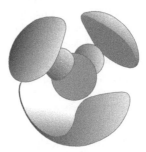

16. **The average first-neighbour environment of a molecule in liquid water.**

The average four-fold motif is clearly reflected in this picture. The two lobes 'capping' the hydrogens of the central water molecule show the range of locations of neighbouring water molecules accepting hydrogen bonds from the central molecule. Curving below the central molecule, rather than there being two separate lobes—one for each of the two waters donating hydrogen bonds to the central molecule—there is a *continuous band* of density.

This reflects the fact that the negatively charged region on the oxygen is a single lobe of charge, which has the structural consequence that water molecules approaching a neighbouring water in the direction of the negatively charged region can comfortably take on a wider variety of orientations than those approaching from the hydrogen side.

The main structural points

The experimental evidence is indeed consistent with the instantaneous structure (V-structure) of water at room temperature and pressure being closely related to that of the ideal random four-coordinated network structure proposed by Bernal. It has the fundamental structural characteristics of a liquid in which there is no long-range order. The local molecular environment of the constituent water molecules and its variability are consistent with the way in which the water molecules interact with each other (the *intermolecular potential*).

On average, the first-neighbour environment is tetrahedral (consistent with the basic four-fold motif), though there are local defects where the coordination can be either greater or less than four. There is a wide distribution of OOO bond angles above and below the ideal tetrahedral angle, and the hydrogen-bonded molecules form a range of ring structures. The ideal random network indeed does seem to be a good reference structure for the real liquid.

Moving around

So far we have focused on picturing the instantaneous structure of water—what the molecular arrangement looks like in a snapshot taken with an exposure time shorter than that in which the molecules themselves can move. But water is a liquid; the molecules are in motion and one instantaneous structure will transform to other instantaneous structures. In detail, these structures will all be different, but each of them will be consistent with the structural properties set out earlier. To get a fuller understanding of water as a liquid, we need to consider how these structures transform from one to another. We need to know something about the dynamical aspects of liquid water.

We are immediately faced with an interesting conundrum. When discussing the difficulty of forming hydrogen-ordered ices, we concluded that the orientational motions needed for the ordering to occur usually required the presence of some kind of defect. This was necessary as the Bernal–Fowler restrictions meant that a single molecule could not move of its own accord without breaking those restrictions.

Our picture of the structure of liquid water is also one of a highly connected hydrogen-bonded network, and so we would expect the liquid molecules to suffer from similar constraints as in the ices. Yet on the melting of ice, there is a million-fold increase in the mobility of the water molecules. This cannot be explained by the breaking of hydrogen bonds on melting as the average number of these per molecule in the liquid is almost the same as in ice. So something else must be giving the molecules in the liquid their ability to move.

Let's look at some numbers. Experimental measurements tell us that at ambient temperature a typical time for a water molecule to *reorient* in the liquid is about 2 picoseconds (or million millionths of a second). Moreover, the mean time it takes for a water molecule to move a distance of about one molecular diameter is around 7 picoseconds. However, the strength of an average hydrogen bond is much greater than the typical energy of the room-temperature thermal fluctuations that would be needed to provide the energy to break it.

These observed water molecule reorientation and translation times are in fact several orders of magnitude shorter than those we would expect if either of these molecular motions were facilitated by the breaking of a hydrogen bond. So some other mechanism must be involved to explain these unexpectedly fast molecular motions in this linked hydrogen-bonded water structure.

There are two obvious possibilities. First, as we mentioned when discussing mobility in ice, it might be possible for a number of connected water molecules to collectively move in such a way that hydrogen bonds are not broken. Several neighbouring molecules might move cooperatively, so avoiding having to pass over significant energy barriers between neighbouring configurations. Another (perhaps related) possibility is to think about a possible role for defects in a similar way to our discussion on rotational mobility in ices. On the basis of computer simulation calculations, it is indeed a structural defect mechanism that is generally regarded as at least one solution to this mobility conundrum.

This view argues that the actual process involves not the charge defects we discussed in Chapter 3 in relation to ices, but the bifurcated hydrogen bond we noted earlier (see Figure 15). Each of these bifurcated hydrogen bonds are estimated to have about half the energy of a single 'normal' hydrogen bond, and the computer simulations that have been performed suggest there can be preferential exchange of molecular neighbours close to these bifurcated defects.

A natural conclusion is that these defects could ease the transition between two locally different hydrogen-bonded configurations, enabling a change in local structure without having to break hydrogen bonds. Thus, although the hydrogen-bonded network in liquid water is relatively robust in not being easily disrupted by normal thermal fluctuations, defect mechanisms appear to give the liquid the mobility we observe.

We thus have a rather interesting picture of the liquid as a relatively robust—dare we say picosecond-time-scale rigid—network of molecules, yet one which, despite the difficulty of breaking the links in that network, behaves as the good, mobile room-temperature fluid that is essential to its chemical and biological functionality.

Heating things up

What happens to this water structure when we increase its temperature? When we heat a 'simple' liquid, whose structure is determined largely by the way its molecules pack together—as exemplified in the random close packing we discussed earlier in this chapter—we expect the average intermolecular distance to increase. Thus the liquid will expand to occupy a greater volume and its density will decrease.

But as we have seen, water is a more open structure: its molecules interact through the basic four-fold motif of Figure 3, which leads to a structure that is more open than that of a simple liquid whose structure is determined largely by packing constraints. So what happens when we increase the temperature of a 'network' type liquid such as water?

Perhaps reflecting the expectation that nothing out of the ordinary happens, not a great deal of experimental work has looked at how the water structure changes as it is heated towards its boiling point. As we would expect, the average intermolecular distance increases, as does the variation of those distances. These changes occur to both the intramolecular O–H bond distances and the first-neighbour (hydrogen-bonded) water–water (O...O) distances. Consequently, the basic four-fold motif will expand.

In addition, we have to consider the effect of increasing temperature on the *geometry* of the basic tetrahedral motif. We would expect there to be an increase in the variability of the OOO angle, and this is what we see. However, we saw in considering the structures of high-pressure ices that an increase in the distortion of this angle increased the ability of the water molecule network to connect itself in different (in the ice case higher-density) ways, so we might expect a similar change in 'intermediate-range' network structure to result from this increase in bond angle distortions. This may be the case, but there are as yet no clear experimental data on the detailed structural changes that actually occur.

Putting on the pressure

For a simple liquid, the application of pressure pushes neighbouring molecules closer together, so reducing the average distance between them. This results in a reduction in occupied volume and an increase in density. For water, we have a similar effect, but the most interesting changes in structure occur as a result of increased hydrogen-bond bending. Remembering that increasing pressure on the ices resulted in increased bond-bending to enable the molecules to occupy smaller volumes (which in the ice case facilitated changes in ring structures), we might expect something similar to occur in the liquid.

What we observe as a result of this increased (OOO) bond-bending on increasing pressure in the liquid is indeed similar. The increase in bond bending allows *second*-neighbour molecules—those that are connected to the central molecule in the motif through hydrogen bonding to the central molecule's hydrogen-bonded first neighbours—to be 'pushed in' closer in to the central water molecule. Though increasingly distorted, the first-neighbour motif remains, retaining on average its four hydrogen-bonded neighbours. It seems to be a remarkably resilient local structure.

We can illustrate this by looking at how the spatial density functions for water (shown for first neighbours under ambient conditions in Figure 16) change with pressure. Figure 17 shows how the first *and second* neighbours are distributed around a central molecule (left) at ambient pressure and (right) in a higher-density liquid produced by increasing the pressure to a few thousand atmospheres.

Although they are partly hidden by the second-neighbour lobes, those representing the first-neighbour distribution are very similar in both images. So the four-fold motif remains. When we look for pressure-induced changes to the spatial distributions of second neighbours by comparing the outer lobes in the ambient and high-pressure liquids, we see both something the same and something different. The second-neighbour lobes look very similar in both cases, but at the higher pressure these lobes have been pushed in closer to the central molecule. Whether there has been some restructuring of the water network as a result of some bond

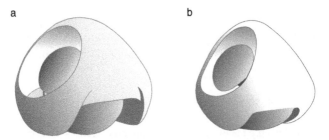

17. **The average distribution of first and second neighbours around a central water molecule at a. ambient and b. high pressure.**

breaking and reforming is not clear, though there are some indications that this might be the case. But it might just be that the structure has adapted to the reduced volume available to it by a uniform compression, an interpretation that seems consistent with the results shown in Figure 17.

We will come across a related discussion later in this chapter when we look at amorphous ices.

Going supercritical

We have seen that increasing temperature and pressure separately puts increasing strain on the directionality of the hydrogen bonding that dominates the structure of water. So what happens when we increase temperature and pressure together? It is well known that supercritical water (above 374°C and about 220 atmospheres pressure, where the density is about one-third of that of water under ambient conditions) behaves very differently chemically from 'normal' water. For example, molecules that are relatively insoluble in water at ambient conditions can become highly soluble, making supercritical water an effective medium for a number of processes such as treatment of hazardous waste and solvent extraction. So it is particularly interesting to try to find out how the structure differs under these more extreme conditions.

Questions we might ask include: how much is the first-neighbour four-fold motif distorted? What are the structural consequences of this? And does the hydrogen-bonding interaction that dominates water structure under normal conditions survive?

With respect to the network structure, the experimental evidence is pretty clear: as temperature is increased above 100°C (but keeping the liquid under increased pressure to prevent evaporation), the water network becomes increasingly distorted. It is largely gone by about 300°C, even though the density may be as

Water as a liquid—and as glass(es)

high as 80 per cent of that of the ambient liquid. The basic tetrahedrally coordinated motif appears to have pretty well vanished.

With respect to how much hydrogen bonding has survived, the position is less clear, as extracting a clear fingerprint for a hydrogen bond from structural data is not easy. However, using what is thought to be a reasonable geometrical criterion for a hydrogen bond suggests that indeed most (though perhaps not quite all) of the hydrogen bonding has been lost by the time the density has fallen to about two-thirds of its ambient value, a conclusion which is in substantial agreement with that from spectroscopic techniques that pick up the presence of hydrogen bonds in a different way.

So it is no surprise that supercritical water behaves very differently from the water we normally experience.

Cool it!

As heating water increases the distortion of the basic four-coordinated hydrogen-bonded motif, we might expect the opposite to happen when we cool water down. Broadly speaking this is indeed the case; the degree of OOO bond-bending falls as temperature falls, and consequently the average local tetrahedral motif becomes less distorted.

Although ice melts on heating and water normally freezes on cooling at 0°C, like other liquids water can be supercooled: under certain conditions its temperature can be reduced below its normal freezing point and still remain liquid. In fact, it can remain liquid down to its *homogeneous nucleation temperature* near −40°C. So obviously we would like to know what happens to its structure as we take the temperature down towards this lower limit: how close might it get to the ideal tetrahedral random network structure?

Unfortunately, it's not easy to make good structural measurements on the supercooled liquid: it's an unstable system that will crystallize if impurities are present. Even when we have removed most of the impurities, it is still difficult to keep the material liquid for long enough to take a good structural measurement.

There are ways of stabilizing the supercooled liquid, for example as very small droplets in emulsions or in very small holes in other materials. However, as those droplets or holes are so small that the water structure itself is likely to be significantly affected by the presence of the surface, we cannot be sure that we are really looking at bulk water. Moreover, there are experimental difficulties in extracting data that are not 'contaminated' by the interactions at the interface between the liquid and the container; for these small droplets the contributions of these interactions to the measured signal can be large and there is no clear way of removing them from the measured data.

However, some neutron and X-ray structural measurements have been taken on bulk water supercooled down to −10°C and very recently on microdroplets right down to the homogeneous nucleation temperature. These show that the average first-neighbour distance reduces as we would expect. Moreover, the degree of hydrogen-bond bending is reduced, resulting in less distortion of the average first-neighbour tetrahedral motif. These network-ordering trends continue as we reduce the temperature further, and by the time we arrive at the homogeneous nucleation temperature near −40°C, we do appear to be getting closer to the ideal tetrahedral network structure.

But *how* close can we get in reality to that ideal structure? What would the structure look like if the supercooled liquid didn't crystallize at about −40°C?

Avoiding crystallization

In many liquids, crystallization on cooling can be avoided by cooling it rapidly. This is the way we produce *glasses*—solid materials that have not crystallized into an ordered arrangement but have retained a disordered structure. This is easier to do for some materials than others—for example ordinary window glass is produced by cooling the silicate liquid quite slowly. But for water, the cooling rate has to be so fast that, although a non-crystalline form of ice was formed by depositing from the vapour on to a cold surface as long ago as 1932, it wasn't until 1982 that water was 'glassified' by rapid cooling of the liquid. And even that work was queried for several years before it became accepted that a non-crystalline, or *amorphous*, form of ice could be made directly from the liquid.

What is the structure of this non-crystalline, amorphous ice material? We know from decades of work that the structure of a silica-based glass is essentially that of its liquid, but with the molecules *frozen in time*; the molecules are fixed in a liquid-like structure. By analogy, we might expect the structure of amorphous ice to be that of supercooled water, with the water molecules forming the same four-coordinated network structure but fixed in those positions.

However, as many of us have learned to our cost, it can be dangerous to draw conclusions too quickly about things related to water. So it might be wise to look first at how amorphous ice can be produced and what experiment tells us about its structure. When we do this, things get rather more interesting than they would have been if we had just made the easy assumption that the water molecules in amorphous ice are fixed in a structural arrangement that is essentially the same as that of the cold liquid.

The first successful preparation of an amorphous solid water in 1932 was through the deposition of warm water vapour on to a very cold surface. Not surprisingly, this material has been christened *amorphous solid water*, or ASW for short. While at first sight we might think that depositing water vapour on to a cold plate is a rather obscure thing to do, and far from having any relevance to real life or to the role of water in just about anything of interest, this form of water is likely to be the most abundant form of solid water in the universe, condensing as it does at extremely low temperatures on to interstellar dust particles. So for its astrophysical relevance alone, perhaps it is worth looking at in more detail. And astrophysicists certainly are very interested in doing so.

We find ASW to be a microporous material highly capable of absorbing gases. The number and sizes of the pores depend on the conditions of preparation, so if we want to study its molecular-level structure, we need to find a way of making a reproducible material. This we can do by annealing samples by heating them to –163°C, a procedure that gives us a reproducible bulk state of ASW with a density (0.94 g/cm³) that is not very different from that of crystalline ice Ih at 0°C (0.92 g/cm³).

But this is only the first chapter of a fascinating—and so far unfinished—story.

Amorphous ice or amorphous ices?

The second chapter in this story starts with the preparation of an amorphous solid form of water by rapidly cooling the liquid, essentially *freezing out* the motions of the molecules so that they don't have enough time to rearrange themselves into their preferred ordered, crystalline, arrangement. Bypassing the crystallization process in this way is easy for many 'good glass-formers' such as the silica-based glasses we might drink our beer from. But water is a particularly bad 'glass-former'. So bad that a cooling rate of over a million degrees a second is needed to prevent it from crystallizing.

It wasn't until the early 1980s that this cooling rate was successfully achieved by one of the foremost ice research groups in an appropriately icy country (Austria), when Erwin Mayer and Peter Brüggeler succeeded by projecting a thin jet of water into a very cold liquid (for example liquid propane at –193°C).

These early attempts, which produced a mixture of amorphous and crystalline ice, caused quite a stir among water experts who had thought that achieving amorphization by liquid cooling would not be possible. As Mayer and his team refined their experimental techniques, solid samples showing no sign of crystalline material were produced that eventually convinced most of the sceptics. So was born the material dubbed *hyperquenched glassy water*, or HGW for short.

Its density was again 0.94 g/cm^3. Although the similar densities of ASW and HGW suggested their structures might be similar, there was no experimental evidence at the time to support or disprove such an assumption.

Soon after the Innsbruck group's work, on the other side of the world in Canada, another leading ice laboratory (led appropriately by a future president of the Canadian Alpine Club, Ted Whalley) muddied the (solidified) waters by producing an apparently amorphous solid water by compressing crystalline ice to above 10,000 atmospheres pressure. Again this may seem an odd experiment to perform, but they had an interesting argument to justify it: if a solid is compressed at a temperature that is low enough to prevent its transformation to another crystalline phase, it might 'melt' to an amorphous solid.

If we go back to the phase diagram of water (Figure 5) and extrapolate the melting line AB to higher pressures as shown, they argued that such a 'melting' transformation might happen if they compressed ice Ih at about –200°C to where it would cross the extrapolated melting line at about 10,000 atmospheres (around point C on Figure 5). They tried it, and indeed the crystalline structure appeared to collapse to an amorphous one. As the density of this material was higher than that of either ASW or HGW, it was christened HDA, or high-density amorphous ice.

Again, the interpretation of this experiment as producing a genuinely amorphous material was controversial at the time and remained so for a couple of decades. Some workers thought that the overpressurization had caused the ice lattice to collapse, and that what was observed was a 'mechanical' melting producing very small 'nanocrystallites' of ice.

So now we have apparently non-crystalline solid water material prepared by three different routes. Does that mean it's time to find their structures and resolve the uncertainties that are building up? Not yet—the story has two more chapters to go before we can do that.

If we take HDA at ambient pressure and heat it slowly to above about –155°C, the material doesn't crystallize, but expands while remaining amorphous. This procedure results in a transformation to yet another apparently amorphous solid form of ice: LDA (*low-density amorphous ice*). Its density of 0.94g/cm³ is similar to the densities of both ASW and HGW. Furthermore, if we keep the pressure on HDA above 8,000 atmospheres and heat it to about –110°C, we get a transformation to yet another apparently non-crystalline form of ice. As a contraction has taken place during this transformation, the density has increased. So this fifth form is, unsurprisingly though rather boringly, called *very high-density amorphous ice*, or VHDA.

Now we'll stop and take a breath before looking at what the structures of these five forms are.

Surveying the (amorphous) ice fields

Table 1 lists these five amorphous forms, together with their methods of preparation and densities. From the latter, it is immediately clear that three of them—ASW, HGW, and LDA—have essentially the same density. Although it isn't necessary for two forms of the same material with the same density to have the same structure, it is quite likely that they would, so the general assumption for many years was that they did. However, it wasn't until high-quality neutron scattering measurements had been made on these materials in the first years of the 21st century that direct structural evidence was obtained to confirm that this assumption was valid: ASW, HGW, and LDA, although prepared by different routes, do have the same molecular-level structure.

Moreover, the water molecules in all three forms seem to be almost perfectly four-fold coordinated as in the ideal four-fold motif, with the mean OOO angle being $\approx 111°$, very close to the ideal tetrahedral angle of $109.47°$. We can thus conclude that the short-range molecular structures of ASW, HGW, and LDA all relate closely to a hydrogen-bonded tetrahedral network of water

Table 1 The five forms of amorphous ice

Form	Acronym	Preparation	Density (g/cm³)
Amorphous solid water	ASW	Water vapour deposition	0.94
Hyperquenched glassy water	HGW	Cooling liquid droplets at 10^7 deg/sec	0.94
Low-density amorphous ice	LDA	Heating HDA to above about −155°C	0.94
High-density amorphous ice	HDA	Compressing ice Ih to above 12,000 atmospheres at −196°C	1.15
Very high-density amorphous ice	VHDA	Heating HDA at about 8,000 atmospheres to around −110°C	1.26

molecules—i.e. to the random network model we have been discussing as an ideal model of liquid water.

On the basis of X-ray scattering measurements on highly supercooled water droplets, we have already commented that as we supercool water, the network tightens up and does indeed appear to be approaching the ideal random network structure. Although we do not have direct evidence to be able to assert it, we might thus postulate that there is some kind of structural continuity between supercooled water and this low-density amorphous form.

The high-density structures

Having now simplified our picture of amorphous ice from five forms to three, what about the structures of HDA and VHDA? As Table 1 shows, these two forms each have a different density which is higher than that of the low-density ASW/HGW/LDA form. What then are these structures? And as they are denser than liquid water, might their structures relate to that of liquid water under pressure that we discussed earlier?

The structures of these two remaining amorphous ices were revealed using neutron scattering measurements, the current interpretation of which says that in both high-density forms, each water molecule is on average connected by hydrogen bonding to four neighbouring molecules. So no change then in the basic local first-neighbour structure: the four-fold motif has survived. However, things look very different if we look at the water molecules that are *just outside* the first-neighbour distance as defined by the four hydrogen-bonded neighbours. For the low-density amorphous ices, there are no water molecules in this region. But on average there is an additional molecule in this region in HDA, and two for the higher-density VHDA.

What appears to have happened is that the need to occupy less volume has resulted in second-neighbour molecules that were originally further away from each central molecule (around 0.45 nm in the case of liquid water and the low-density amorphous ices) have been forced closer to that central molecule. They are now between 0.31 and 0.35 nm away from it, but without being directly hydrogen-bonded to it. Although the pictures are not quite identical, the spatial distributions of the first and second shell neighbours look similar to those of liquid water under pressure (Figure 17). So it appears that the high-pressure amorphous ices may have structures similar to those of liquid water under pressure.

Amorphous polymorphism

Before we found out that we could produce more than one structurally distinct form of amorphous ice, there was a general feeling that there was likely to be only one amorphous structure for a given substance. This is unlike the case of crystals where—as we have seen in spades for ice—molecules can often be arranged in different regular ways to form different crystal structures.

This phenomenon of *polymorphism* can be particularly interesting, for different crystal forms can have different physical properties. This point is certainly not lost on, for example, the pharmaceutical industry, who may need to ensure a drug is in a particular crystal form to optimize uptake by the body. For amorphous forms, however, it was thought that with the crystal constraint being absent, there would be only one way of arranging molecules in a non-crystalline way. With these results on ice, we have been forced to consider the possibility of *polyamorphism*, or *amorphous polymorphism*.

So in the tradition of being sceptical scientists, perhaps we should consider the possibility that these amorphous ice structures are not distinctly different. We saw earlier that putting liquid water under pressure modifies its structure by the non-bonded second neighbours being pushed closer in; but we didn't see then a need to call this an essentially different liquid structure.

On the basis of our discussion of the high-density amorphous structures above, doesn't the structural relationship between the near-random-network LDA structure and the structures of the high-density amorphous ices seem similar to that between low- and high-density water? In which case, should we really think of the low- and high-density amorphous ices as having distinctly different structures?

Although the evidence for or against such a proposal is not conclusive, there are quite convincing data that suggest we are indeed dealing with three distinct amorphous ice structures. Careful measurements show that there is a large jump in density in going from LDA to HDA: if these two forms had essentially the same underlying structure we would have expected to be able to move smoothly from one to the other. Similarly, when transforming HDA to VHDA, although there isn't a large density change, there is a significant change in compressibility. Thus the evidence at present does imply that we have a genuine case of amorphous polymorphism.

Also relevant to this argument is the observation that, in HDA and VHDA, there is quite a lot of distortion of the local tetrahedral motifs. We discovered when discussing the higher-pressure ices in Chapter 3 that the stress from highly distorted hydrogen bonds can be reduced if some hydrogen-bond linkages 'thread through' rings formed by other molecules (see Figure 7). Might not something similar be happening here, with some ring threading occurring in the high-pressure amorphs?

If this is the case, then we would expect a discontinuity between a non-threaded and threaded structure, a topological transition that would likely be mirrored in a discontinuity of some physical properties—and as we noted earlier there is a change in compressibility in going from HDA to VHDA.

However, the stress relief resulting from the threading would likely result in reduced distortion of the basic four-coordinated tetrahedral motif, and this is the opposite of what is observed experimentally. Moreover, attempts so far to detect threading in computer-simulated amorphous ice structures have failed to find any.

So while the jury is still out, the weight of evidence to date is that we should regard LDA, HDA, and VHDA as distinct polyamorphic forms.

Are amorphous ices truly non-crystalline?

Before concluding this amorphous ice discussion, we should note one further point on which there is still some disagreement: are these amorphous forms genuinely amorphous or might they be just linked assemblies of very small crystallites? Or, noting the large number of different crystalline ice phases, may they not be just mixtures of small crystallites of different phases?

In trying to resolve these issues, serious attempts have been made to interpret the neutron scattering data in terms of crystallinity. However, all the microcrystalline models that have been tested so far have failed to agree with the experimental data. Thus, although it is difficult to be *absolutely sure* from neutron (or X-ray) scattering data that a sample is genuinely non-crystalline, the majority opinion is that these ice structures are indeed genuinely amorphous.

Chapter 5
The anomalies explained

Much is made of the fact that water sometimes behaves differently from most other liquids. This anomalous behaviour has been recognized for very many years, so people have wondered for a long time why water seems to behave oddly. In a remarkable lecture to the Nottingham Literary and Philosophical Society in 1887, James M. Wilson, who was then a mathematics master at Rugby school but was later to become Archdeacon of Manchester, suggested that all the anomalies might well depend on a single property of water. Was he right? Let's try to find out.

Some of the so-called anomalies we have already come across. For example, whereas most other hydrides (for example hydrogen sulfide) are gases at normal temperatures and pressures, water is a liquid. Moreover, water remains liquid over a relatively wide temperature range (100 degrees Celsius) and it takes an unexpectedly large amount of energy to vapourize it at its boiling point—its latent heat of vapourization is rather high.

We explained both these apparently anomalous properties in Chapter 2. They are simple consequences of the strength of a typical hydrogen bond between water molecules. As we learned earlier, at room temperature this energy is equivalent to several times that of a typical thermal fluctuation that would be needed to break it. In contrast, most other small molecules interact mainly through the much weaker van der Waals forces. As the typical energy of this weaker interaction is generally significantly less than that of thermal fluctuations at room temperature, most small molecules are unable to remain in the liquid state under ambient conditions.

Useful oddness

Not quite so simple to explain are what we might call the *structural* anomalies of liquid water. For example, liquid water is denser than ice—in stark contrast to most other systems where the liquid phase is less dense than the crystalline phase. Moreover, cold water *contracts* on warming, with its density passing through a maximum at 4°C. Consequently, not only does ice float on water, but also, because very cold water is less dense than slightly warmer water, the colder water will tend to sit on the surface of a lake. Both these unusual properties have major consequences. For example, cold lakes stratify. They also freeze from the top rather than the bottom, with the non-convecting and low-conductivity floating ice layer slowing further freezing. This is quite important if you're a fish.

Both of these properties of water thus are of climatological and biological importance. Indeed, it is often hypothesized that the overall biological importance of water is specifically related to its anomalous properties. Whether this is so or not remains a matter of argument (see Chapter 6). What is not a matter of argument are the molecular reasons for the so-called anomalous behaviour.

What's normal and why?

Simply stated, when a 'normal' liquid freezes, its volume decreases. Conversely, when a 'normal' crystal melts the density falls—the liquid occupies a greater volume per molecule than does the crystal. When we heat the liquid up, it expands. This makes it easier to compress and so the compressibility increases as we increase the temperature. When we put pressure on a liquid, its viscosity increases, and the ability of the molecules to diffuse in the liquid—its diffusivity—falls.

We can understand why this is normal liquid behaviour by referring to the random packing model of a simple liquid that we discussed in Chapter 4. Look at Figure 12, which compares the ordered arrangement of spheres in a crystal with the disordered one of an ideal liquid. The density of the disordered packing material (the ideal liquid) turns out to be about 15 per cent lower than that of the ordered crystalline packing; hence the liquid must be less dense than a crystal of the same material close to its melting point. This is a basic geometrical fact: we cannot pack spheres more efficiently than in the regular crystalline packing shown in the figure. Though this has been known experimentally to be the case for over a hundred years, proving it mathematically is extremely difficult; it is only recently that this conjecture has been proved.

If we now heat up our 'normal' liquid (or indeed a crystal of the material), it is common knowledge that it will expand as the average distance between pairs of atoms increases. If we now increase the pressure, we will squeeze the volume the liquid occupies, and hence reduce the space which the molecules need in order to be able to move as well as increase the strength of the forces between the neighbouring atoms. Consequently, the diffusivity will fall and the viscosity will increase.

Water behaving badly

Figure 18 shows schematically (top) how the volume of water and (bottom) its compressibility changes with temperature. Shown for comparison is the expected behaviour of each property for a typical 'normal' liquid.

We see immediately that water does things quite differently. At lower temperatures, each of the properties shown does the 'wrong' thing.

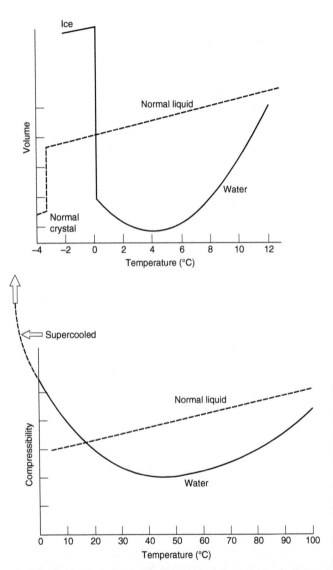

18. How temperature affects the volume (top) and the compressibility (bottom) of water, compared to what is observed for 'normal' liquids.

80

Ice *contracts* on melting; a normal liquid expands. Below 4°C liquid water *contracts* on heating; a normal liquid expands throughout its temperature range. Between the melting point and 46°C water's compressibility *falls* as temperature increases; for a normal liquid it increases. If we look at water's viscosity at or below about 30°C, it *falls* as we increase pressure from 1 up to about 1,000 atmospheres; it increases with pressure for a normal liquid.

Water apparently is behaving badly at low temperatures. However, as we heat it up it seems to come into line with 'normal' behaviour. The reason for this apparently schizophrenic character (abnormal at low temperatures, normal at higher ones) relates to the four-coordinated local structure that is the basic structural motif in water. This bad behaviour is all down to the local molecular geometry that reflects the directionality of the hydrogen-bond interaction (Figure 3).

The melting anomaly

For the normal liquid idealized in Figure 12, removing the crystallinity constraint on the hard sphere crystal allows the molecules to occupy a larger volume. The liquid is therefore less dense than the crystal. For water, however, we have already seen that the structure of the normal crystalline phase, hexagonal ice Ih, is expensive on space—Figure 4 shows the open channels that are a consequence of the 'open' tetrahedral geometry of the water coordination in our basic four-fold motif. On melting ice, we remove the crystallinity constraint. The molecules are then able to explore a wider range of local geometries, and as we discussed in Chapter 4, a range of different instantaneous local structures results (see for example Figure 15).

If we look at this melting in terms of ring structures, ice Ih contains only six-fold rings. Releasing the crystallinity constraint enables the liquid to form a range of ring structures. These will include five- and four-fold rings, both of which will take up less space per molecule than the six-fold rings of ice Ih.

Alternatively, we might look at the volume change in terms of the distortion of the basic four-coordinated motif. We know that in ice Ih, these motifs are pretty close to the ideal tetrahedral geometry. In order to fit into the reduced volume available to them in the higher-pressure ices, these basic local molecular units have to distort by increased bending of the OOO angles, a distortion which helped the formation of a range of different ring structures. Put slightly differently, the more perfect the four-fold motif, the more volume it occupies. Conversely, as we go to structures in which this basic motif is more disordered—in this case melt the crystal—the system can occupy less volume. So its density is increased.

Looked at either way—as a change in distribution of ring structures or distortion in the liquid of the underlying four-coordinated motif—the removal of the crystallinity constraint allows the molecules to take up a range of local structures, many of which occupy less volume per molecule than the crystal. The overall result—as is perhaps implied in the two-dimensional cartoon of Figure 14—is liquid of a *higher* density than the crystal. Hence ice contracts on melting.

Which means water expands on freezing. And in doing so can do a fair amount of damage to systems as widely different as biological cells and rocks. So this anomaly is rather important both geologically and biologically—as well as perhaps giving a clue as to why food from the freezer might not taste as good as before it was put into cold storage.

Cold and schizophrenic

We can make a related argument to explain the low-temperature thermal expansion behaviour shown in Figure 18. As we know, a simple liquid expands on heating as the average first-neighbour separation increases. This happens in water also; but in addition, increasing the temperature increasingly distorts the OOO angles in the basic four-fold motif. The consequent increase in the bond angle variation will (as it did on the melting of ice) enable the exploration of *denser* local structures.

Thus we have two effects operating as we increase the temperature—a normal thermal expansion process resulting in increasing the average intermolecular hydrogen-bond distances and a *reduction* in density, and an increasing angular variation that will lead to an *increase* in density. These two effects thus work against each other, one trying to increase the volume, the other to decrease it. The volume change we observe will be the result of the resolution of these two competing tendencies. At low temperature (below the temperature of maximum density of 4°C) the bond angle variation dominates, leading to an overall volume contraction; above this temperature the normal thermal expansion mechanism takes over and the liquid begins to expand more like a normal liquid.

We can explain the anomalous low-temperature compressibility behaviour in a similar way. As we see in Figure 18, in contrast to a normal liquid, cold water becomes more difficult to compress as we raise the temperature until we get to 46°C. Above that temperature its behaviour returns to normal. Again we have a competition between two opposing effects. At lower temperatures, applying pressure will increase the average distortions of the basic four-fold motifs through further bending of the OOO angles. This will allow denser local structures to form, and the consequent pushing closer together of the water molecules will make it more difficult to compress the liquid. Once we have passed the minimum in the compressibility curve, the 'normal' thermal expansion process dominates, reducing the liquid density and therefore making it easier to compress it.

We can also explain in a similar way the anomalies in dynamical properties such as viscosity and diffusivity. In normal liquids an increase in pressure reduces the specific volume per molecule which will reduce the ability of the molecules to move; hence a reduced diffusivity. Cold water does the opposite—applying pressure increases it diffusivity. The explanation is essentially the same as discussed earlier: the pressure increase will distort the local orientational order through increased hydrogen-bond bending. Remembering how we invoked the presence of bifurcated defect structures in Chapter 4 to explain the mobility of the liquid water hydrogen-bonded network, increasing the pressure will likely increase the defect concentration, and thus increased diffusivity will be expected. Viscosity is simply related to the inverse of diffusivity, so a similar explanation serves for the anomalous behaviour of the viscosity.

Is this odd-ball behaviour unique?

The simple answer is no. If our explanations of these anomalous properties are valid, wouldn't we expect other materials whose structures have a similar local tetrahedral symmetry to also behave oddly? We would. And they do.

We are very familiar indeed with one of them—silica. In SiO_2, each silicon is connected to four oxygens in a four-coordinated motif very similar to that we have been considering for water, and each oxygen is shared between two neighbouring silicons. So it should be no surprise that we do find similarities to water in the properties of its liquid. For example, it has a temperature of maximum density that can be explained in the same kind of way that we have explained why water passes through a maximum density at a particular temperature. Other examples include silicon and germanium, which ideally bond to their neighbours with a tetrahedral local geometry.

Moreover, some other atoms or molecules that link together through directed bonding may have open structures, and so show related 'anomalous' behaviour: for example bismuth also contracts on melting. We might note that although this behaviour is similar to that of water, the stronger intermolecular bonding in these materials means that this behaviour occurs at much higher temperatures: compared to the 0°C for water, the melting points of SiO_2, Si, Ge, and Bi are respectively 1710°C, 1410°C, 937°C, and 271°C.

So our expectations are correct. It *is* the local intermolecular geometry that is ultimately responsible for the structural 'anomalies' that can be observed not only in liquid water, but also in other liquids with similar local geometrical structures. It looks like we should posthumously congratulate J. M. Wilson for his foresight in Nottingham in 1887 in suggesting that a single structural property may be responsible for water's unusual properties. Moreover, despite frequent claims to the contrary, there is no real 'mystery' about the origin of water's physical properties. We can understand them in terms of the directionality of its intermolecular interactions. And in these terms, water is not unique.

The universal solvent?

There are further aspects of the behaviour of water that, while not considered anomalous, make it particularly interesting, as well as extremely useful and important chemically, physically, biologically, geologically, and environmentally. These relate to the electrical properties of the molecule. Although these are intimately connected with its molecular structure, it is simpler to look at these properties separately from the structural anomalies.

From our discussion of the molecule itself in Chapter 2, we know that it possesses separated regions of positive and negative charge. The simplest picture we showed in Figure 1 had notional positive charges on the hydrogens and a negatively charged region on the opposite side of the oxygen. As we know unlike charges attract, a simple model of the hydrogen-bonding interaction between two water molecules is for the positive regions to attract the negative ones, resulting in our familiar four-fold local structure.

Water is, however, far from being alone in being a *polar* molecule—one in which there are separated charges on the molecule. Therefore, if we mix water with other polar molecules, we would expect like–unlike attractions to occur also with the added molecules. The result: an aqueous solution of the added molecules. Similarly, if we added a salt to water, it would also dissolve easily: the positive ion of the salt would happily seek out the negative region of the water, while the negative ion would go for the water's positive hydrogen end. Water thus is particularly good at dissolving other polar or charged molecules or ions.

It is not unique in this ability to dissolve 'stuff' but it is particularly good at it. This is for two main reasons. First, as we mentioned in Chapter 2, its dipole moment, which quantifies the strength of a molecule's polar quality, is quite high. However, as we also mentioned in Chapter 2, in addition water has a high dipole polarizability: its charge distribution is relatively easily increased when sitting in an electric field.

In the liquid, each water molecule will experience a (fluctuating) electric field from its neighbours. This field will perturb its electrical charge distribution. Overall the effect in the liquid will be to increase the average molecular dipole moment in the liquid. Although there are no easily interpretable measurements of the degree of enhancement of the dipole moment, calculations based on realistic liquid structures suggest a wide distribution of individual dipole moments results, ranging from *c.* 2 to *c.* 4 Debye, with an average therefore significantly higher than the isolated molecule value of 1.85 Debye that we gave in Chapter 2.

Breaking up is *not* hard to do

A result of this dipole moment enhancement is to further increase the power of liquid water as a solvent. Let's consider for example adding some common salt (NaCl) to water. The effective charges on the sodium and chloride ions are larger than those on the water molecules, so why wouldn't the sodium and chloride ions prefer to stick together in solution—after all, the energy that would bind them together is larger than that which would persuade them to be promiscuous and pair up with a water molecule or two and so dissolve? The reason is that water has a rather nasty (if you believe in monogamy) trick up its sleeve in that its very presence weakens the attraction between the oppositely charged ions.

This is due to the basic physics of the interaction between two charges. We know that the force attracting two charges q^+ and q^- depends on both the sizes of the charges and their distance apart. The interaction weakens the further apart the ions are—the well-known inverse square law.

However, the force is modulated by the nature of the medium in which the interaction takes place, and the more polar the medium, the weaker the interaction. This weakening is quantified by the *dielectric constant* or electrical *permittivity* ε: the larger the ε, the weaker the interaction. And liquids of molecules with a high dipole moment have large permittivities. So the attractions between the ions are broken up and the water molecules cluster round the ions which have now dissolved into the solution.

Ions beware! Water is a very powerful solvent.

Shifting the charges

In addition to molecular diffusion being anomalously high in liquid water, so also is its electrical conductivity. At first sight this seems odd—why should a collection of neutral H_2O molecules conduct electricity at all? We already have a clue to the answer from our discussion of defects in ice.

Though we have assumed so far that liquid water is made up of neutral H_2O molecules, it's not quite that simple. This is lucky for us, as if there were only neutral molecules then lots of things important to our existence wouldn't happen. The joker in this pack is that water molecules can dissociate into ions though a dissociation reaction:

$$2H_2O \Leftrightarrow H_3O^+ + OH^-.$$

These ions—or hydrated versions of them such as $H_5O_2^+$—are present in only very low concentrations: about one pair of ions in around ten million water molecules under ambient conditions. The fact that they are there, however, means that water is both a weak acid (presence of positive ions) and a weak base (presence of negative ions). The balance between the concentrations of positive and negative ions—and hence the acidity or alkalinity—can be shifted by adding other molecules. Changing temperature also affects the equilibrium of the dissociation—at 1,000°C and 100 atmospheres pressure around one in every thousand waters is ionized, making the liquid a very different chemical beast.

The presence of even the normal very low concentration of ions in liquid water is sufficient to enable charge transfer processes—and hence electrical conductivity—to take place. And as we shall see later in Chapter 6, the ability to move charge around is important when considering the biological importance of water.

The conduction mechanism

Although we know these ions are involved in conducting charge through water, there is still controversy about the actual mechanism. For many years, the favourite explanation was a process suggested over two centuries ago by the German chemist Theodor Grotthuss, who has been credited with establishing the first theory of electrolysis and in formulating the first law of photochemistry.

As illustrated in Figure 19, a proton from an H_3O^+ ion is argued to move along a hydrogen bond to a neighbouring water, so recreating the H_3O^+ on the neighbouring molecule. Another

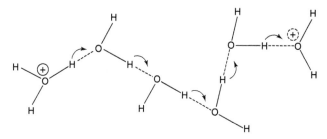

19. Theodor Grotthuss's idea of how positive electrical charge might be conducted in water.

proton from the receiving molecule then translocates similarly to another neighbour, etc., etc., resulting in the original 'excess' proton moving through the liquid.

A similar process can be invoked to try to explain the 'reverse' movement of an OH⁻ ion—try it by drawing a similar linked hydrogen-bonded chain of waters with an OH⁻ at one end, and then allow a hydrogen from its neighbour to turn it into H_2O. Then continue along the chain until you reach the other end where you will have created a new OH⁻ ion. The negative charge will thus have been transferred from one end of the chain to the other.

Attractive as this mechanism appears, it's in conflict with some of the experimental evidence.

An explanation that does seem consistent with experimental data is illustrated in the series of structural snapshots in Figure 20. Molecule A in the top picture is an H_3O^+ ion. As it has three hydrogens it hydrogen bonds to three other water molecules through directing each of those hydrogens towards the negative regions of three neighbouring waters.

In order for the additional proton to be able to move along a hydrogen bond, the receiving neighbour in A's first coordination

20. How we now think protons move in water.

shell (molecule B) has to lose one of its four neighbours—say molecule C, which is in the *second* coordination shell of the H_3O^+ (molecule A). The breaking of this hydrogen-bond (see the middle picture) is thought to control the proton diffusion rate. Once the proton has shifted to the neighbour B, molecule A is short of a fourth neighbour so another water molecule D hydrogen bonds to it.

The process can thus be thought of as being propelled by hydrogen-bond cleavage in front of the moving proton with subsequent hydrogen-bond formation at its back. An analogy might be Moses parting the Red Sea.

Enter quantum mechanics?

It has been fashionable for at least forty years to think that protons may *tunnel* along water molecule chains. In this well-known quantum mechanical effect, a particle that doesn't have enough energy to get over an energy barrier may in effect be able to tunnel 'through' it. Attractive as the idea may seem, recent calculations do not support the idea of this mechanism as being of significant importance here.

However, what calculations do show is that the barrier to proton motion along a hydrogen bond is 'washed out' to some extent by the zero-point motion of the hydrogens that we referred to in Chapter 2. This zero-point motion is thought to effectively reduce the classical energy barrier to proton motion by about 75 per cent, so easing the proton transfer process. As zero-point motion is greater the lighter the atom involved, the fact that water has two light atoms is important for this charge transfer in the liquid. And hence for its anomalously high proton conductivity which, as we will see in Chapter 6, is likely to have significant biological relevance.

Chapter 6
Water as a biomolecule

Water is a biomolecule

Water is essential for life—so we are frequently told. All known life forms contain a pretty high fraction of water: whereas around 70 per cent of each of us is water, a jellyfish beats us tentacles down with a water content of around 98 per cent. So it's hardly surprising that we have come to believe that water is a defining parameter for life. If there is no water, we are told, there can be no life.

This mantra extends to our search for extraterrestrial life, which seems to concentrate on finding places where water might be (or might have been) present. Looking for water on Mars has been a preoccupation for decades. And on the face of it, there is plenty of evidence that the familiar forms of life certainly do not thrive in the absence of water. For example, drying food has long been an effective means of preserving it from microbial attack.

On Earth, of course, there is no other ubiquitous naturally occurring liquid available, so the evolution on Earth of a life form that depended on a liquid medium other than water is hardly likely. If the life we see on Earth is dependent on water, is this simply an inevitable consequence of the widespread presence of liquid water on Earth? Or is it because water might be *uniquely* capable of supporting the existence of *anything* we would recognize as living?

Given this recognition of the biological importance of water, it is perhaps rather surprising that biology teaching tends to recognize the so-called 'molecules of life' as, for example, the enzymes that catalyse life-depending chemical reactions, the nucleic acids that carry the genetic code, and the lipids that form the cells in which many of the life-sustaining operations take place. Little if any consideration is generally given to the aqueous medium in which many biological processes take place.

We might recall here why Bernal wanted to understand liquids (Chapter 4): he wanted to understand water and its role in biological processes. If life does indeed depend on water, then the water molecule itself should perhaps be regarded as a biological molecule in its own right as much as a protein or a nucleic acid. Perhaps more so if water is indeed essential to the operation of these larger, more romantic biomolecules.

Why is water biologically important?

If indeed *all* life depends on water, we ought to be able to say *why* it does. As more scientists have begun to realize the biological importance of water, recent years have seen an increasing number of scientific conferences and workshops on 'water in biology'. Yet these rarely seem to ask precisely *why* water is important. What biological functions does water take part in at the molecular level? Is the water really necessary? Is there something unique about the properties of water that means it cannot be replaced by another compound?

If we cannot answer these questions, how can we be sure that water is *essential* to life? Could other molecules fill a similar role or roles? Could water be replaced by some other medium in some other life form elsewhere in the universe that can still metabolize and reproduce?

We will raise briefly these last two questions towards the end of this chapter. For now, however, we will dig down into our knowledge of water and its interactions to see if we can answer the two perhaps less difficult questions: *how* does water influence biological processes and *what* properties of the water molecule enable it to do so.

Stabilizing protein structures

As the range of biologically relevant processes we might consider is vast, we'll focus on the role of water in the structure and operation of enzymes to illustrate the ways in which water is involved.

Enzymes are large protein molecules, which are made up of a 'string' of smaller (amino acid) molecules that are linked together chemically. For the enzyme to do its job in catalysing a chemical interaction such as the chemical modification of another molecule—the *substrate* on which it acts—it must first take up a structure that enables it to work. The extended string of amino acids needs to *fold up* to form a more compact arrangement in which the parts of the molecule that are active in catalysing the chemical process are exposed so that the substrate can access them.

A simple way of thinking about the enzyme–substrate interaction is that it involves a *lock and key* mechanism: the substrate (the key) has to be able to attach itself to the active part on the enzyme (the lock) so that the chemical interaction (turning the key) can occur. As we know from experience, if a lock is damaged, the levers inside it are not quite in the right place and the key will not turn. Similarly, if the enzyme does not have the correct structure, the substrate may not bind, or even if it does, the chemical process may not take place.

Thus, the *structure* of the enzyme is critical to its activity. So it is natural to focus first on the *structure* of the macromolecule and see how water may be relevant to both forming and maintaining that structure.

The universal solvent again

To do this, we obviously need to consider how water molecules interact with other molecules.

We have already done the spadework here in Chapter 5, where we noted how water molecules, being strongly polar, can interact through hydrogen bonding with other polar molecules. So as many of the chemical groups on the amino acids that make up our enzyme are polar, for example containing alcoholic –OH and amidic –NH groups, we would expect water to interact easily with them. And we would expect these interactions with other polar groups to occur in ways that are consistent with the geometry of the interaction. Remember again our four-fold motif!

When we look at the structures of enzyme molecules, we find that they tend to fold up into ball-like compact structures (not necessarily spherical—rugby balls are a closer analogy for some enzymes). The surfaces however are not necessarily smooth at the molecular level, and there may well be pockets, particularly at active sites where the substrate needs to bind (a characteristic that is perhaps one reason for the lock and key analogy). In folding to this compact structure, various chemical groups of the amino acids will interact with each other—for example through hydrogen bonds similar to those we have considered in the case of water-water interaction.

However, at the surface there will be chemical groups exposed. If these exposed groups are capable of hydrogen bonding to other groups, then unless they can find a polar group with which to interact, the structure is likely to be energetically unstable with respect to the unfolded structure, and so the macromolecule may not fold.

It is here that water comes to the rescue.

Satisfying mutual needs

Figure 21 illustrates the kinds of interactions that water molecules make with a typical biomolecular surface. This example shows a section through a pocket on the surface of vitamin B_{12} coenzyme, a molecule that is essential for the activity of several enzymes and is involved in a number of biochemical processes. In fact, vitamin B_{12} was one of the structures for which Dorothy Hodgkin was awarded the 1964 Nobel Prize in Chemistry, and it was with her strong encouragement that this work on the solvent structure was undertaken.

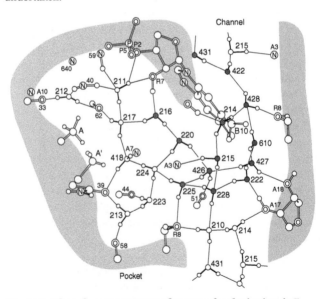

21. **A snapshot of an arrangement of water molecules in vitamin B_{12} coenzyme.**

Some of the water molecules, for example those numbered 217 and 224, seem to be very happy, linking to neighbours by four hydrogen bonds as in the four-fold motif—two hydrogen donor and two hydrogen acceptor interactions in each case. And looking at the water molecule labelled 213, we see that it donates both its hydrogens to the exposed (polar) oxygens of the groups on the coenzyme labelled 39 and 58, as we would expect it to. Similarly, water 211 hydrogen bonds through its negative oxygen end to (again polar) hydrogens on the coenzyme groups labelled 40 and 59. So these water molecules seem to be interacting with their (water and non-water) neighbours as we might expect them to interact.

However, if we are a bit more critical and look more closely at these last two water molecules, we find that things aren't quite as we might expect. Water 213 hydrogen bonds to only one other hydrogen (from water 223). And the 'upper' hydrogen of water 211 seems to interact directly with two, not one, polar groups on the coenzyme surface. Thus our hydrogen bonding is not perfect, just as it is not perfect in the liquid. We might also note in passing that the bond angles and distances we observe in this apparently complicated water network can be rationalized in terms of the repulsive restraints put on the intermolecular interactions by the actual shape of the molecule that we noted in Figure 2—Bernal's 'impenetrability' that was key to the structure of simple liquids also plays a significant structural role in water.

If we measure the structure of water in these biologically relevant systems, we find that it is essentially unchanged from that of the bulk liquid. The water molecules hydrogen-bond with the polar groups of other molecules in ways that are consistent with the way its molecules interact in its own liquid. This includes adapting to three- and five-fold coordination where (respectively) insufficient or excess hydrogens are available to make up the ideal four-fold motif.

So water seems to be very comfortable in these complex environments. In this coenzyme B_{12} example, we see that it can fill 'crevices' in larger molecules in an energy-efficient way. And as the structure of the macromolecule is essential for its biological operation, the ability of water to 'mop up' interactions with polar groups both in crevices and on the exposed external surface that, if not satisfied, would help to destabilize the enzyme, enables it to stabilize that structure and hence enable its biochemical activity.

Put simply, water molecules are critical in maintaining the active structures of other biomolecules. And the versatility of the water molecule in its interaction with other molecules (remember the sixteen phases of ice in Chapter 3 and defects such as bifurcated hydrogen bonds in Chapter 4) would seem to be a critical property that enables it to perform this structure-stabilizing role.

'Hydrophobia'

Looking again at Figure 21, we might notice that not all the coenzyme surface groups shown interact directly with the surrounding water. For example, the two groups towards the left of the figure labelled A and A' are methyl ($-CH_3$) groups. These are *non-polar* groups: they do not have a significant separation of charge that would help them interact directly with another polar group such as water. Therefore, they have little attraction for water molecules. So even when such a non-polar group is nearby, water has little energetic incentive to interact with it. It is much more comfortable (i.e. it is more energetically favourable) for it to avoid the non-polar group and hydrogen-bond to a nearby polar group. For example, as we see in Figure 21, water 212 prefers to link to the polar oxygen of the group labelled 33.

This non-polar-group avoidance results here in a network of water molecules (assisted by a polar oxygen from group 62) passing around these two methyl groups from the oxygen of 33 to that of group 39. There is no strong interaction between the waters forming this partial 'cage' and the non-polar groups A and A' on the coenzyme's surface. Similarly, a (much larger) cage is seen around the benzimidazole group at the top of the figure as shown by the water molecules with shaded-in oxygens.

Why is this 'caging' important? First, it shows us that, even where non-polar groups are present on a biomolecule surface, the water can still arrange itself in a 'comfortable' (or energetically acceptable) structure. The presence of non-polar groups will therefore not necessarily destabilize the biomolecule's structure in water.

An idealized representation of the water structure around the non-polar methane molecule is shown in Figure 22; we see that the water molecules arrange themselves into a spherical cage-like structure around the methane. Because the methane is non-polar, it cannot hydrogen-bond directly to the surrounding water molecules, which themselves link to each other through hydrogen bonds.

In this structure, each water molecule retains four neighbours. Three of them are shown in the figure within the shell, with the fourth connecting to water molecules elsewhere in the structure. In terms of the detailed geometry of the cage, we see five-fold rings that are consistent with the near-tetrahedral local geometry of our familiar four-fold motif. And as it prefers, each water molecule donates two and accepts two hydrogen bonds from its neighbours.

So even when there is no hydrogen-bonding possibility with a non-polar molecule or group, water can accommodate the non-polar group's presence by this kind of caging.

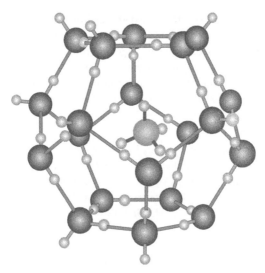

22. The 'cage' arrangement of water molecules in a crystal of methane hydrate.

Plugging the gas pipes

The kind of structure shown in Figure 22 is commonly known as a *clathrate* structure, clathrate being derived from the Latin *clatratus* meaning 'with bars' or 'a lattice'. Depending on the nature and size of the molecule in the cage, there are a number of different crystalline clathrate structures that water can form, though all are consistent with its preferred near-tetrahedral interaction geometry. These can be a particular problem in oil and gas recovery, with the result that much research has gone into finding ways of preventing their formation in pipelines.

Moreover, large amounts of methane naturally frozen as clathrates exist both in permafrost formations and deep in the ocean. As methane is a much more potent greenhouse gas than carbon dioxide, the potential release of this methane, for example as the permafrost melts, will have very significant implications for climate change. And massive release of methane from deep below has been put forward as an explanation for the loss of ships in the Bermuda Triangle, defined loosely as a triangle with the tip of Florida, Bermuda, and Puerto Rico as its vertices. However, although there are large methane hydrate deposits off the south-eastern United States, no large releases that could cause a ship to sink are believed to have occurred for the past 15,000 years.

Although these crystalline clathrate structures have long been thought to be stable only when there are 'guest' molecules occupying a significant fraction of the cages, it has now been shown to be possible, with great care, to pump out small guest molecules to leave the empty cage structure. Such structures are the ices XVI and XVII that we met in Chapter 3, the future study of which could help our understanding of the stability and properties of these important but sometimes troublesome structures.

The hydrophobic interaction

Pleasing as these cages may look, they are not just 'pretty faces'. Quite importantly, they lead us to considering what many have thought for over fifty years to be a major biologically relevant phenomenon linked with water. This is the so-called *hydrophobic interaction*.

This is usually described as the tendency of two non-polar molecules or groups to 'attract' each other in water. Look again at our cage structure in Figure 22. If we now bring another similar cage close to it, the two methane molecules will come into contact, with a consequent rearrangement of the surrounding water molecules to form a larger cage around the contacting pair of methanes. The water molecules will still be perfectly happy in terms of their hydrogen bonding to each other—remember their structural versatility—but as the surface area of the resulting cage will be less than that of the two separate cages, not all the original 'cage' water molecules will participate in the new, larger cage. Those that do not will be freed from the cages, and expelled to the surrounding liquid.

The classical explanation of this apparent attraction between non-polar groups in water rests on an assumption that the water molecules participating in the surrounding cages are in some way *more ordered* than those in the liquid. If this is so, then as bringing together two non-polar groups will result in the expulsion of some water molecules from the cages to the liquid, there will be an overall decrease in the order in the system. Such a decrease in order is energetically favourable, and so encourages this coming together of non-polar entities.

Although we needn't go into technical details, in thermodynamic language the freeing of some of the original cage waters into the bulk will result in an increase in a disorder-related property of the system called *entropy*. As reducing the free energy of a system is a driving force for a (physical or chemical) process, and an increase in entropy will contribute to such a loss, this bringing together of non-polar groups in water is considered to be an *entropy-driven* process.

And the experimental evidence?

I should come clean here, however, and note that this explanation—though accepted for several decades—is coming under increasing questioning. When put to the test by neutron scattering measurements, no structural difference between the water in the cages and that in the bulk liquid is detected. Experimental evidence for any structural enhancement of water near a non-polar entity is indeed hard to come by.

Moreover, measurements of actual structures in solutions of non-polar entities demonstrate that in the liquid, the cage structures themselves are not as idealized as the picture in

Figure 22—they are much more disordered than that picture suggests. A much simpler explanation that is consistent with experiment exists: it may be no more than a consequence of incomplete mixing of the two components. We might also note in this context that this kind of interaction in solution also occurs with a range of other non-aqueous solvents. This general *solvophobic* effect can also be explained in a similar way, without any need to invoke solvent restructuring.

Whatever the explanation is of this tendency for non-polar entities to come together, it is a real effect that is of biological importance.

A three-dimensional jigsaw puzzle

We have now described two ways in which the molecular groups that make up a folded protein can interact in an aqueous environment. Polar groups in the protein might hydrogen-bond to other polar groups, or to a surrounding water molecule. Consequent on the hydrophobic interaction (irrespective of what its physical explanation is), non-polar groups might prefer to be in contact with other non-polar groups in the interior of the protein, or if on the surface, surrounded by a cage-type network of waters. In addition, where charged groups exist on a protein, these could interact either with oppositely charged groups or with water molecules. And in some proteins which include sulfur-containing amino acids, there can be internal covalent links through disulfide S–S bonds.

There are thus a number of different kinds of molecular interactions that can occur when a protein folds up into its active, or *native*, structure. In order to be stable with respect to its unfolded *denatured* structure, these different kinds of interaction need to be largely satisfied through making appropriate liaisons. So folding the protein might be thought of as doing a three-dimensional jigsaw puzzle. But a rather complicated one in which there are different kinds of pieces that can fit together only with specific (similar or different) kinds of pieces. Polar pieces fit with other polar ones (in the right directions); non-polar ones with other non-polar ones; positive charges with negative ones; sulfurs with other sulfurs. But for the native protein to be stable, all these different kinds of interactions must be largely satisfied.

For a complex large macromolecule, this looks like a tall order—the active structure needs to satisfy a large number of different constraints. Here water seems to come to the rescue. It fits comfortably in available hydrogen-bonding sites. Water molecules can accommodate irregular interfaces and can make hydrogen bonds where no other polar group on the biomolecule is available to do so. And through three- and five-fold coordination, which is an important aspect of its structural versatility, it can mop up donor/acceptor imbalances, for example where a single hydrogen 'bifurcates' (as does that on water 211 in Figure 21) to help satisfy the needs of two separate oxygens that have no other polar hydrogen option available.

So the integrity of the active protein structure is highly dependent on water. Not only because of the hydrophobic interaction which has been touted since the 1950s as the major driving force in protein folding, but because of its versatility in satisfying interactions that otherwise would not be possible. Promiscuous perhaps, but pretty neat!

We can follow a similar argument when looking at interactions between macromolecules and other molecules. When a substrate binds to the active site of an enzyme, water molecules also can be important in ensuring that the hydrogen bonding is satisfied in the enzyme-substrate complex. When two macromolecules interact, they also need to come together in such a way as to satisfy the intermolecular bonding requirements of the groups on the surfaces which come into contact. For a macromolecule whose exposed surface is a mixture of polar and non-polar groups (as is the case with most globular proteins), we would expect non-polar groups on one molecule to interact with non-polar groups on its neighbour. We would also expect exposed polar groups on one molecule to hydrogen-bond with polar groups on the neighbour. And in cases where the polar groups are not close enough or not properly aligned for good hydrogen bonding, an intervening water molecule or two can be brought in to make the link remotely.

Bringing two completed jigsaw puzzles together is thus another jigsaw puzzle of the same kind.

Biological compartments

Much biological activity takes place in cells—compartments that are filled with many different molecules. Their walls not only separate the contents of the cell from its external environment, but also control the passage of selected molecules to and from the outside. Through the hydrophobic interaction, these walls depend for their integrity on water.

The main constituent of animal cell walls are *phospholipids*, long chain molecules with a strongly polar head group and a non-polar hydrocarbon chain tail. As we would expect, the polar head groups will interact directly with water, while the non-polar chains will, as a result of the hydrophobic interaction, prefer to interact with each other. So we can understand how a combination of these interactions will lead to the formation of a *bilayer* structure as illustrated in Figure 23.

The role of water in forming and maintaining these structures is central. And it is also implicated in mechanisms by which other molecules pass through the membrane in order to fulfil their biological function.

Some other contents of water's bio-portfolio

The biological roles for water discussed so far are largely static ones relating to its function in biopolymer structure and stability.

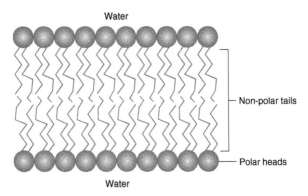

Water

Non-polar tails

Polar heads

Water

23. A schematic of the way lipid molecules are organized in a bilayer membrane.

It can also take a more active role—for example, there is good evidence that water is an active participant in the proton pump mechanism that transports protons across the cell membrane. It is also implicated as an active participant (for example as an acid or a base) in some of the chemical reactions that are catalyzed by enzymes.

There is also a role for water as a transport medium. In order for biology to work, molecules have to move around from place to place. So they need an appropriate fluid medium in which to do so. Water obviously plays such a role and there are good reasons why it does the job well.

First, an effective biological medium must be a good solvent—and we have already explained why water is one. Its dielectric permittivity makes it a good solvent of charged and polar molecules. And it is also able to accommodate non-polar groups in solution.

Second, for a substrate molecule to get to the active site of an enzyme, it has to be in a medium through which it can move. It may seem a bit obvious to say that water is indeed an appropriate liquid medium, but as we discovered in Chapter 4 when discussing molecular mobility, the relatively strong hydrogen-bond interaction implies that water may not be a particularly good medium for other molecules to move through. However, as we found earlier, the existence of transient bonding defects ('bifurcated bonds') seems to provide a mechanism whereby, even though its liquid network is relatively robust on the picosecond time scale, it still behaves as a good, mobile room-temperature fluid that is needed for chemical, and therefore biological, functionality.

It is not only the ability to enable molecules to move that is important for an effective biological medium. In many biochemical reactions, charges (protons or electrons) have to move. For example, a proton may need to get to a site on a protein for whose action the proton is chemically necessary. Again, water seems to work here. As we discussed in Chapter 5, protons are able to move with relative ease through a hydrogen-bonded chain of water molecules, though we may not yet be absolutely sure of the reasons why they can. Although proton conduction may not be relevant to the particular coenzyme B_{12} system shown in Figure 21, a potential conducting path of water molecules can be seen from the surface water molecules 215 and 431 into the pocket.

Third, we have so far completely neglected the possible influence that the dynamics of the water molecules may have on biological processes. This is an area where there is much discussion and a number of different views (some of which are looked at again in Chapter 7). But there can be no real doubt that, as the aqueous environment is a liquid, the motions of the molecules in the liquid must be linked in some way to the motion of the biopolymers, be they structural proteins, enzymes, polynucleic acids such as DNA, cell membranes, or whatever.

Here I need to make yet another admission. Up to now, I have (again!) been guilty of painting a far too static picture of things. Looking at the coenzyme B_{12} in Figure 21, you would be excused for thinking that the water molecules are static—that they are fixed in the positions shown. *This is not the case.* They are in constant movement, with a given molecule being continually replaced by another molecule. What is shown in Figure 21 is a time-averaged picture that helps us to understand the issues of structural importance. But everything—the biomacromolecule as well as the solvent molecules—is in constant motion. And as the water molecules interact with the exposed groups on the biomolecule, then the motions of the biomolecule and the water must somehow be linked.

How this dynamic linking influences biological activity remains largely an open question. There is some evidence that the dynamics of the solvent may actually drive the dynamics—and hence the activity—of the biomolecule: the biomolecule would then be the slave of the solvent rather than the other way round as has generally been assumed. Even if this master–slave relationship is not the case, it is highly likely that the dynamics of the solvent molecules will modulate that of the biomolecule and hence influence its activity.

However, whatever the solvent, there would likely be some mutual influence between solvent and biomolecule. This is consistent with the observation that where an aqueous solvent is replaced by a largely non-aqueous one, the activity of the biomolecule can change—and quite dramatically in some cases. Though how much of this should be put at the door of the changed dynamics remains unclear—all sorts of other things will have changed, for example the dielectric permittivity and perhaps some small changes in enzyme structure.

But perhaps we can reasonably argue that, having evolved in an aqueous environment, the biomolecule's activity will have been optimized for that environment.

What properties are biologically important?

Can we now draw out some promising molecular-level characteristics that may begin to answer this question?

We have repeatedly come across the underlying ideal tetrahedral arrangement of the local interaction geometry. It was the key to understanding the so-called anomalies in Chapter 5. This geometry does appear also to be central to the structural versatility of the molecule that we have argued above is significant for its biological importance in stabilizing protein structures. Moreover, the asymmetry of charge distribution, with the positively charged end of the molecule (the hydrogen donor) being more orientationally constraining than is the negatively charged region (the hydrogen acceptor), also contributes to this structural versatility.

This asymmetry makes both trigonal and tetrahedral local structures 'comfortable' and enables hydrogen bonding to be satisfied even in situations where there is a mismatch between the number of available donors and acceptors in a complex macromolecular structure. So not only is this tetrahedral geometry—four hydrogen-bonded neighbours made up of two proton donors and two proton acceptors—biologically important for maintaining the stability of native enzyme structures; so also may be its asymmetric nature. Other 'more perfect' tetrahedrally coordinated systems such as silica, silicon, or germanium may therefore lack an important characteristic that water possesses when it comes to being a 'sympathetic' and useful biosolvent.

Second, we noted in Chapter 4 that, with a hydrogen-bond energy being about ten times that of a typical room-temperature fluctuation, we would expect the mobility of a water molecule in the liquid to be much less than it is found to be. The existence of a significant population of local coordination defects—in particular bifurcated hydrogen bonds—was there argued to explain the observed molecular mobility. We thus appear to have a relatively strong molecular framework, yet one that has a much higher mobility than we would expect it to have. Again, the reason for this seems to relate to the details of the imperfect tetrahedral geometry. So if biomolecular processes require a relatively stable solvent framework with an inbuilt ability to allow relatively rapid molecular movements, then water seems to have put itself in pole position to do the job effectively.

Third, liquid water is a good proton conductor, and proton conduction is necessary for certain biomolecular processes. The conduction process seems to require the relatively resilient framework we have just commented on, but one which is again relatively labile at the individual molecular level. Moreover, as we noted in Chapter 5, the proton transfer step itself may be eased by zero-point motions reducing the height of the energy barrier the translocating proton has to cross.

Whether or not these characteristics are essential to any biomolecular process that might be built on a different chemistry is of course arguable. But with respect to the way life has developed in water, a case might be made for it having developed so as to exploit these particular characteristics, in addition to the more obvious ones like the solution properties of a high dielectric constant liquid. One might envisage other molecules which have some of the above characteristics. But can we think of one, apart from water, which has them *all*?

In *The Hitchhikers Guide to the Galaxy*, the supercomputer Deep Thought was set the task of working out the 'Answer to the Ultimate Question of Life, the Universe, and Everything'. It found it a time-consuming job, even for a supercomputer. It finally came back with the answer: '42'. I like to think of this as meaning four hydrogen bonds, with two donors and acceptors. Perhaps this core property of water *is* the meaning of life.

Biosolvent engineering

One way of exploring why water is important to biological processes is to replace it with another medium. Another is to take the water away from an operating biological system. Observing what happens in either case, or noting the properties of the replacement medium in the former situation, might throw some further light on the aspects of water that are indeed biologically important.

Roy Daniel, a long-time biochemist New Zealand collaborator of mine, helpfully assigned water in enzymatic systems into four categories.

First, we need a diffusion medium for the substrate to get to the active site and the reaction product to move on to where it is needed. For us, this medium is liquid and aqueous, but that doesn't mean it need necessarily be so. Second, there is the water surrounding the active enzyme, with the inner aqueous shell in direct contact with the enzyme surface (such as in the coenzyme B_{12} example of Figure 21). Third, there can be water molecules inside the biomolecule that hydrogen-bond to internal polar groups and that therefore contribute intimately to the stability of the active structure. And finally, where the chemical reaction catalysed by the enzyme involves water specifically (e.g. dehydration or hydrolysis), water is obviously a prerequisite.

Starting with water in the first and second categories, we can replace most of it by some (not all!) other organic molecules with the enzyme remaining active. From this we can conclude that the active structure is not destroyed. However, the activity is usually significantly reduced, though by choosing the solvent appropriately, it is possible to engineer enzyme systems that are *more* active than they are in an aqueous environment.

One reason given for reduced activity is that, in media of lower dielectric constant (activity has been observed where the dielectric constant of the medium is forty times lower than water), the enzyme is less flexible, with its mobility correlating with the dielectric constant of the medium. This reduced mobility is likely to at least contribute to the lowering of activity—for example the molecular movements involved in the catalytic process are likely to be slower. If we now start slowly adding water molecules back, step-by-step, the flexibility of the enzyme begins to increase. And, unsurprisingly, so does the enzyme's activity.

We can thus conclude that as we increase the degree of hydration of a system in which much of the water was initially replaced by organic solvent, both enzyme activity and flexibility increase. However, the enzyme is still able to function—albeit rather slowly—even when most of the surrounding medium is a 'suitable' organic liquid. So in a completely water-free environment, enzyme activity may not be prohibited by the absence of water per se, provided the solvent permits a catalytically competent conformation to be maintained. In such cases, water may not be necessary.

Completely drying up

What happens to an enzyme's activity when we completely remove the water? How is its structure affected? We have argued that the water (or a suitable organic liquid medium) is important in stabilizing the catalytically active structure of an enzyme. Therefore, we might expect that taking away the water will cause the loss of that structure, resulting in an inactive enzyme.

Interestingly, this need not be the case. Just as substituting water by organic solvent reduces the biomolecule's mobility, so does removing the water. And as the enzyme needs to be sufficiently mobile to change its structure, dehydrating it is likely to make it more difficult for it to denature. So we shouldn't be surprised if it *doesn't* lose its overall catalytically competent structure on dehydration, especially if the water is removed by freeze drying, a process in which the water is pumped off after the preparation has been frozen and the enzyme's mobility further reduced.

We can identify which water molecules come away most easily on dehydration, and follow how the removal of certain waters (for example those hydrogen-bonded to amidic or alcoholic groups on the enzyme) affects both the chemistry (the processes required for effective catalysis) and physics (the flexibility) of the enzyme. It is, however, very difficult to remove all the waters. In particular, any water molecules that are internal to the folded protein (those in Roy Daniel's third category mentioned earlier) will be particularly difficult to budge. If we do manage to remove them, we would likely destroy the native, active structure.

Assuming we can remove the water in the first two categories (the surrounding fluid and the waters in contact with the enzyme surface), if we now want to measure residual activity we have a bit of a problem: we no longer have a liquid medium through which the substrate and product can diffuse respectively to and from the enzyme's active site.

Roy Daniel's group resolved this neatly by working with gaseous, rather than liquid, substrates and products. By this clever device, they separated the product and substrate diffusion role of water from any specific water requirement of the catalytic process. These results interestingly concluded that, in the case of the particular enzyme system chosen, there is no absolute requirement for water in that particular enzyme's catalytic action. Neither the flexibility of the enzyme that water enhances, nor the surface hydration water, was essential for activity. The activity may have been low, but the beast was still active. So what can we conclude from this?

Life as we *don't* know it?

These experiments on enzyme activity in water-free preparations demonstrate that there is at least one biologically important process that we thought needed water that does not in reality require it. Although we cannot generalize from one example of an enzyme that can function anhydrously, it does imply that some kind of anhydrous life process *might* work—although envisaging how such a system might originate anhydrously is another question.

However, we have also seen that as we add water to a dry enzyme, activity increases, and of course the rates of enzymatic processes are important in maintaining life processes. And as the biological processes on which we depend have evolved in an aqueous medium, the rates of these processes are tuned to that medium. It is one thing to get an enzyme to work in a non-aqueous environment, or in a non-aqueous organic medium. It is a very different thing to get 100,000 different and connected processes to work in concert—as they have to do to support our existence.

While this underlines the biological importance of water to life 'as we know it', this discussion does perhaps suggest that, had a different solvent been present on Earth (or elsewhere in the universe), the properties of that different solvent might have been exploited in the development of some different form of life. There have been interesting discussions of the kinds of characteristics such a solvent might need to have to enable the molecular synthesis, metabolism, and reproduction processes that characterize living things. Some of these discussions are listed in the 'Further reading' section. They do suggest that, in our search for life elsewhere in the universe, we might fail to find it if we focus entirely on the presence of water as a prerequisite for life.

Chapter 7
Some past and current controversies

What's left to understand?

I hope it's reasonably clear by now that the structure, dynamics, and properties of water are in the main pretty well understood. We have a good idea of its structure, and how that structure explains why its properties—some of which are relevant to its roles in maintaining earthly life—are sometimes different from those of a 'normal' liquid. However, there remain a number of issues that continue to raise their heads that suggest that there may still be some things about water that we don't yet understand.

In the popular media, we get offered specially treated water that we are told has wonderful powers to cure us of various ills or keep us extra healthy. Water is thus given some kind of 'magical' status that we would, it seems, be silly not to avail ourselves of. But also in the serious scientific literature, interesting controversies have arisen. Some of these have been effectively knocked on the head by good experimental work. Others remain to be resolved.

These debatable issues are of two kinds. The first relate to characteristics that have been assigned to water that are inconsistent with either our current understanding of water itself or, more seriously, conflict with well-established physical principles. The second kind break no known laws of physics but have yet to be resolved experimentally. We'll discuss examples from each category, starting with one of the really persistent ones.

The 'memory' of water

This old chestnut is the basis of various claims for water, particularly its possible therapeutic value over and above our basic biological need for it. It comes in many different forms. For example, it has been claimed that if we put good words or intentions into it, we can transfer to it 'love, gratitude, and healing vibrations'. Others claim that by immersing plants in water and exposing to sunlight, the 'healing energies' of plants and flowers can be transferred to the water. By doing something to water, we apparently can give to it some kind of property that remains with it in some form—it has a memory. We are told we should choose the kind of water that will best serve us.

Knowing what we do about water, it is quite easy to reject such claims as fantastic. However, the same kind of concept is inherent in homeopathy, a system of alternative medicine developed over 200 years ago and still used today. (Controversially, you can still get homeopathic treatment in some areas of the UK on the National Health Service.)

It is based on the suggestion that *like cures like*: a substance that causes the symptoms of a disease can also cure similar symptoms in sick people. However, recognizing that significant amounts of such substances cause disease, the substance in question is dissolved in water to reduce its concentration. It is then repeatedly further diluted, with the greater dilutions thought to be the more powerful in treatment. Although some practitioners use dilutions that do contain some molecules of the original substance, the 'end product' is often diluted so much that it contains not a single molecule of the original substance. So any efficacy the preparation might have must depend on the water retaining some kind of memory of what was originally dissolved in it.

Many scientific papers have been written on the memory of water. One of these was published in a major scientific journal, even after peer review by the five independent experts consulted recommended rejection for scientific reasons. There have also been times in my scientific career when I have been extremely frustrated by an experiment involving water not working as my understanding of the physics said it should do. Frustrated so much that in the darkest hours (admittedly sometimes in the middle of the night during a long experiment) I've thought that some sort of memory effect must be the explanation.

One particular case involved trying to form ice III. According to the water/ice phase diagram (see Figure 5), I should be able to do this by pressurizing ice Ih to about 3,000 atmospheres at about −50°C to form ice II. I can then warm ice II which should transform to ice III at about −30°C. It didn't. It remained as ice II. So we warmed it up further, passing through the region of stability of ice III without seeing any sign of the expected stable phase until the ice melted.

We scratched our heads. What was happening? Perhaps the transition from ice II to ice III is rather slow and we didn't give it enough time? OK, no problem—we should just cool our now-liquid sample and it will recrystallize to ice III. It didn't. We got ice II again. We repeated the melting and cooling process several times, but always with the same result. So as ice II was the crystal form with which we started, it was tempting to think that the water retained a memory of its previous crystalline phase so that it preferred to crystallize back to it. A desperate hypothesis for a sane physicist to have to make. But in the end, after further thought and experiment, we found a perfectly good explanation of what we had observed that did not require postulating a long-term memory for water—and incidentally led us to a new discovery that we hadn't anticipated.

In fact, water *does* have a memory. But it is a *short-term* one; a *very* short-term one of the order of a few picoseconds—a few million millionths of a second. As we learned in Chapter 4, a typical time for a water molecule to reorient is about 2 picoseconds at ambient temperature. Furthermore, the mean time it takes for a water molecule to move a distance of about one molecular diameter is about 7 picoseconds. So the basic physics tells us that there is no way that a sample of liquid water can remember for more than a few picoseconds how it might have had its structure changed by a previously dissolved substance. Put another way, if some structuring of water induced by a substance is active in curing a particular symptom, the shelf life of the 'medicine' is a few million millionths of a second.

Polywater

In the mid-1960s, a researcher in the Urals claimed to have produced a form of water that was denser and more stable than ordinary water. His small samples were produced in narrow quartz capillary tubes held over a bath of water, from which water evaporated and then condensed as the 'new' liquid in the capillary tubes. Knowing that because of its basic tetrahedral local structure the density of water is relatively low, it seemed plausible that there might be another way for the water molecules to arrange themselves with a higher density. We do, after all, have amorphous ice structures with densities higher than that of water (Chapter 4), so why not a higher-density liquid phase?

The problem was the claim that the purported new form was more stable than ordinary water. If it really was, then there would be a danger that every single water molecule on Earth would transform into this more stable form and, incidentally, wipe out life in one fell swoop. Water as a weapon of mass destruction?

This work might not have gone any further had not it been taken up by Boris Derjaguin, one of the 20th century's most eminent scientists who laid much of the foundations of modern surface science. He improved the production technique, and measured a number of its properties. The freezing point was around −40°C or less and the boiling point around 150°C or more. Its high viscosity meant it flowed more like syrup than water, and the density was some 10–20 per cent higher than that of normal water—though it wasn't recognized at the time, this is about the same as the high-density amorphous ice we discussed in Chapter 4.

He presented this work at a conference in Nottingham in the UK in 1966. Had it not been presented by such an eminent scientist, it might have died a natural death. But the claimed findings, advocated by a recognized and much respected scientific world leader, were so strange that the race to find the structure of what was then called merely *anomalous* water began.

One day in late 1966, I walked over to a hotel near Birkbeck College, London, to pick up Boris Derjaguin for a discussion with Bernal, with whom I was doing my PhD on liquid structure. Derjaguin brought with him a small box containing a capillary tube of anomalous water and presented it to Bernal. A very interesting discussion followed, as a result of which I and my colleague Ian Cherry (and Paul Barnes who joined us later on the project) were assigned the task of sorting out what this stuff was.

We tried hard to make it, placing the water bath and capillary tubes in sealed desiccators, from which the air was pumped out in an attempt to speed up the formation process. We failed—much to Bernal's frustration. Elsewhere, people jumped on the bandwagon, some apparently driven by the prospects of a Nobel Prize and big bucks. The US Navy were interested in it for reasons that were classified and that we therefore found rather interesting to speculate about. A theoretical chemistry group in the US performed high-quality quantum mechanical calculations and concluded that the substance was in fact a polymer: the water molecules bonded to each other through covalent chemical bonds rather than the weaker hydrogen bonds in normal water. Hence the coining of the term *polywater* $(H_2O)_n$.

Back at Birkbeck, we continued to have problems. So we tried other ways of producing the material, still without success. We were beginning to look like incompetent experimenters. Finally, almost in desperation, we took out the vapour trap in the vacuum line between the desiccator and the vacuum pump and...bingo! There were columns of liquid in the tubes. So what now?

Although we were impatient to try to find out how the water molecules were arranged in this material, the first thing we needed to do was to confirm that we were dealing with pure $(H_2O)_n$. So we sent some of our 'best' samples to colleagues for electron microprobe and mass spectrographic analysis. The results came back: what we had was oil from the vacuum pump, which immediately explained the ease of production once we had removed the bit of kit that was designed to prevent such contamination of the samples.

Others were coming to similar conclusions. We had pump oil. Others had silica dissolved from the inside of the tubes, while another worker found similarities with the sweat he personally produced after a game of handball. What you got depended on how you tried to make the sample and the precautions you took—or didn't take. Eventually, the build-up of experimental evidence led Derjaguin to retract the original work, accepting that 'these [anomalous] properties should be attributed to impurities rather than to the existence of polymeric water molecules', though a few years earlier, following our *Nature* paper on 'Polywater and Polypollutants', he had written to Bernal to say we should be sacked for incompetence.

So we could breathe again. No longer need we worry about all the water in the world transforming itself to a more stable—and likely biologically inactive, if not toxic—form.

Surface ordering

It is widely accepted that, close to a flat, solid surface, the ways in which the molecules can arrange themselves must be restricted by the very presence of that surface. We can see this in Bernal's hard sphere model of a simple liquid: as shown in Figure 12, the molecules tend to line up in semi-ordered layers close to the surface.

A similar ordering effect might be expected for water close to a flat surface. There we would expect the water molecules to be arranged in a slightly different—some would say *more ordered*—way. If so, the properties of the interfacial water might be at least a little different from those of the bulk water. And considering that much of the water in a biological cell will be close to some kind of surface, if interfacial water does behave differently from the bulk, then that could have significant implications for the role of water in biological processes. In exploring this issue, it's useful to consider separately how (a) the structure and (b) the dynamics of the water might be affected by the presence of a 'typical' surface.

How far from a surface does this ordering go? There is considerable experimental evidence that has been interpreted as showing that the effect of the surface extends not just to one or two molecular layers but much further out into the bulk liquid. If this is indeed the case, then we would expect the properties of the aqueous phase some distance from the surface to be significantly perturbed from that of the bulk. This would have implications for the mobility of biologically relevant molecules such as enzyme substrate molecules, for example.

Considering that the cell contains a large number of differentiated structures each with their own functions (e.g. ribosomes, mitochondria), and is indeed pretty crowded with 'stuff', the further this 'ordering' propagates, the more we would expect the properties of the water in the cell to be affected. In fact, some workers argue that the properties of *all* the water in the cell are perturbed from those of the bulk liquid. So much so that it behaves as a viscous liquid. If this is so, then there would be a major effect on both the mobility of molecules within the cell and the interactions between the different cell constituents.

However, we should remember that we are not dealing with uniform, flat surfaces at the molecular level. Moreover, the surfaces tend to be chemically complex, exposing a mixture of polar, charged, and non-polar groups to the external medium. One thing we have learned from high-resolution crystallographic studies of proteins is that the water is so structurally versatile that it accommodates itself quite comfortably to this geometrical and chemical complexity. In those interfacial situations, the structure of the liquid at the surface looks pretty similar to that in the bulk. So if there is any structural 'ordering' of the liquid, it would appear to be rather limited.

If the water in the cell is indeed more viscous than the pure bulk liquid, then we would expect this to show up in measurements of the dynamics of water molecules close to biological molecules. But here also, the experimental evidence suggests otherwise.

It can be argued that the clearest and least-challengeable data on the dynamics of water molecules in complex systems come from nuclear magnetic resonance measurements. This is not a simple technique when it comes to interpreting the results of the measurements—in fact, in the 1970s the conventional interpretation was that the motions of water molecules close to a protein were slowed down by around a million times. As the technique developed, however, this conclusion was seen to be highly dependent on the model used to interpret the data, and more sophisticated work began to demonstrate contradictions within the interpretative model used.

To cut a long and tortuous story short, the current picture is much simpler. There are a few water molecules in isolated pockets within some proteins where the motional retardation is large—several orders of magnitude compared to bulk water. These water molecules are really constituent parts of the protein structure itself; they are the ones we mentioned in Chapter 6 that cannot be removed without the protein denaturing. As they are isolated within the macromolecule, they cannot influence processes occurring at or near the protein surface.

In stark contrast, the vast majority of the water molecules in the hydration layer of protein molecules are slowed down by a factor of only around two compared to bulk water. The mobility of these surface waters therefore remains high. Consequently, the thermally activated processes that take place at the interface between the protein and the aqueous phase (such as substrate binding, recognition, and catalysis) can take place at reasonable rates. We are still working in a liquid, not a viscous syrup.

So there is little from these results that suggests there is a major influence on enzyme activity from either structural ordering of water molecules or the slowing down of their motions near a biological interface. However, strong arguments remain, especially regarding the state of water in the crowded environment within the cell.

Two liquids?

Although this particular hare starting running in 1992, its origin goes back to the middle of the 20th century. Then, our understanding of water structure was much more rudimentary than it is now, and was dominated by models that have long since been superseded in the light of experimental evidence.

For example, an early model that was used to fit the X-ray diffraction data envisaged liquid water as a disordered ice-like structure, with additional 'interstitial' molecules occupying some of the large holes in the ice Ih structure. Another model due to Linus Pauling was also based upon the idea of interstitials in a disordered lattice structure. However, his lattice was not that of an ice but of a clathrate, with water molecules occupying some of the cavities that are occupied by guest gas molecules in the gas hydrates (see Chapter 6).

Other 'chemical' models treated water as an equilibrium mixture of different 'species'. In some of the earliest of these models, the 'species' were small aggregates of water molecules, for example, $(H_2O)_2$ ('dihydrol') and $(H_2O)_3$ ('trihydrol'), which mixed with single H_2O molecules to form the liquid. Another popular mixture model of the 1960s was the 'broken hydrogen-bond' model. This defined the different 'species' in terms of H_2O molecules that formed 0, 1, 2, 3, and 4 hydrogen bonds with neighbouring molecules.

Although some of these models continued to be developed in the 1960s, their earlier versions were criticized in Bernal and Fowler's classical 1933 paper (see Chapters 3 and 4) as being 'conceived too much in the manner of molecular chemistry'. They pointed out that this kind of model gave an inadequate description of the actual structure of the liquid in terms of the relative arrangement of molecules. As we have seen in Chapter 4, Bernal preferred to lay the foundation of the random network model that we now see as an ideal model of the actual liquid structure.

If we look a bit more deeply into the idea of mixture models, we can see why they were a tempting route to follow in trying to understand the apparently anomalous properties of liquid water. Take the fact that water (anomalously) contracts on heating up to its 4°C temperature of maximum density, after which it expands as we would expect for a normal liquid. In Chapter 5 we explained this behaviour in terms of two opposing tendencies. One of these (bond bending) enables the density to increase as the structure is heated, while the other (bond stretching) decreases it. Below the temperature of maximum density, the volume contraction mechanism dominates, while above it the volume expansion process takes over.

In contrast, a simple mixture model could explain this behaviour, not in terms of opposing tendencies as temperature is increased, but rather as a change in the equilibrium populations of two different kinds of water structure. These different kinds of structure might be conceived of in a number of ways, but in the simplest form of the model the liquid is seen to be made up of *dense* and *bulky* forms. Although many variants have been proposed, in most of these *two-state models* the bulky species was assumed to be clusters of hydrogen-bonded molecules (perhaps distorted 'ice-like', perhaps resembling the ideal four-coordinated random network), while the dense species was considered to be more closely packed (remember the ideal random packing model of a simple liquid) and of higher energy.

The properties of the liquid are then explained in terms of the equilibrium between the dense and bulky species, an equilibrium which shifts as the temperature or pressure is changed. With appropriate fitting of parameters, some of the experimental thermodynamic data can be reproduced. For example, changing the temperature will change the relative populations of the two different-density species, with the result that the density can pass through a maximum.

Though later experimental measurements, complemented by the growth of computer modelling, demonstrated these mixture models to be both unnecessary and in conflict with experimental data, the idea of two possible liquid structures of different density arose in a different and more subtle guise in 1992. Although, as we have seen in Chapter 5, there are alternative ways of explaining the anomalies we have so far discussed, the behaviour of other properties of water with temperature led to new postulates which are rather interesting and intellectually challenging.

We can illustrate this with respect to the compressibility of water, which we discussed in Chapter 5 as being anomalous. As we saw in Figure 18, it passes through a minimum at 46°C. What I didn't point out in Chapter 5 was that as we reduce the temperature in the supercooled region, there is a dramatic rise which gets steeper and steeper (see Figure 18). As this is the kind of behaviour we would expect when approaching a critical point, at which the distinction between two phases (e.g. between a liquid and a gas) appears to vanish, we would dearly love to see what happens as we continue to decrease the temperature.

No luck. Murphy's Law rules! We hit the homogeneous nucleation temperature at about −40°C and the supercooled liquid crystallizes.

There are competing theories on the reasons for this rapid rise in compressibility (as well as rapid changes in other thermodynamic quantities). The most controversial of these is that we may be approaching a critical point of two liquids with different structures. In the light of the data, this is a reasonable enough hypothesis, so we should be able to test it by experiment. But Murphy's Law again gets in the way as the postulated critical point would appear to be found in deeply supercooled conditions that we just cannot access. The liquid insists on crystallizing before we can get to it. If this liquid–liquid critical point is there at all, it's in a 'no man's land' that we just cannot get to experimentally.

Before summarizing the current state of play on this two-liquid hypothesis, it's a good idea to step back a little and think about what is being suggested. At first sight, the idea of a liquid having two different structures with a definite transition of phase between them seems odd. Liquids are fluid systems, in which the molecules are in continual movement, changing their positions and neighbours all the time. Thus we expect there to be *continuous* small changes in overall structure of the liquid (for example, as temperature or pressure is changed, as discussed in Chapter 4), rather than there being a definite, discontinuous change of liquid structure as we get to a particular temperature and pressure.

However, there are good theoretical arguments to support such a possibility. Provided the intermolecular potential function—the way in which the molecules interact with each other—is of a certain form, two distinct liquid structures are a possibility. There are also experimental data that suggest such a two-liquid scenario may have been observed in some liquids. So why not in water? Particularly when we already know there are at least two forms of amorphous ice at ambient pressure: LDA and HDA. Perhaps these are the (ideal) structures of these two liquids? Some years ago, this was thought to be a reasonable suggestion—until VHDA was discovered, when some were quick to adopt that as the 'reference' structure for the higher-density liquid.

As we cannot access experimentally the postulated critical region, perhaps we can get to it by computer simulation, where we can control the system in such a way that we can prevent it crystallizing?

Much effort has indeed gone in to such calculations since the 1990s. Some of the early ones even claimed to find three or more different liquid structures for simulated water, but things seem to have settled down to trying to find good evidence for the existence of just two differently structured liquids. There are considerable technical problems in not only computing phase boundaries, but also in identifying different liquid structures in simulations. Much progress has been made in tackling these issues, but the two-liquid situation for water remains unresolved.

Even if computer simulations did demonstrate the existence of two different liquids for a particular computational model of computer simulated water, we could really only conclude that two liquids existed for that particular potential function. It is still the case that we do not know the water potential function sufficiently accurately to be confident we are simulating something akin to real water. So answering unambiguously the question of whether there is or is not a liquid–liquid transition in real rather than computer-simulated water will need some very imaginative experiments that somehow get round the problem of the inaccessibility of no man's land. Experiments using laser heating of micron-sized droplets have tried to do this, but the conclusions remain controversial.

This hare looks like continuing to run for some time yet.

Conclusions?

The uncertainties remaining about water, some of which have been discussed here, suggest we should indeed have a question mark in this heading. However, this should not detract from the fact that the structure and behaviour of water itself is pretty well understood.

Although it can be argued that the foundations were laid by Bernal and Fowler on that fogged-in night at that Moscow airport in the 1930s, our ideas on water structure have evolved through chemical, mixture, disordered crystal, and interstitial models to our present picture related to random four-coordinated networks. This is consistent with the experimental data that tell us about liquid structure. In particular, the neutron and X-ray scattering measurements that have been made possible by developments in both these sources and detector systems, have been instrumental in getting to our current level of understanding. Robust interpretation of these data has been further improved through the imaginative use of high-power computing resources. And, interestingly, some of the better potential functions used in computer simulations of aqueous systems—be they of pure water or complex biologically important macromolecular systems—relate very strongly to the simple water molecule model that Bernal and Fowler put forward in their influential 1933 paper.

When comparing them with other liquids, some of the properties of water are seen to be anomalous. These anomalies do have relevance to the importance of water in critical areas such as influencing our climate and in the operation of biological systems. We can explain these anomalies in principle in terms of the underlying local liquid structure. We may perhaps also be able to explain them through a two-liquid scenario. But as such a picture doesn't seem to be necessary to understand the anomalies, we might ask if catching that particular hare—which in itself would be a scientific *tour de force*—will significantly increase our knowledge of why water behaves as it does.

Beware the hype!

In John Polkinghorne's book *Quantum Theory: A Very Short Introduction*, his final section is headed 'Quantum hype'. In that section, he issues an intellectual health warning against using the strange and surprising nature of quantum theory to explain things we don't understand like telepathy and consciousness. I'm tempted to issue a similar health warning here by cautioning against 'water hype'. Yes—water is a very interesting and unusual liquid which is necessary for life, but that doesn't mean it has to have magical qualities, or the mysterious curative or psychologically enhancing properties that are variously claimed for it.

But scientifically it really is a fascinating system. Studying it in its various guises is incredibly stimulating. It can on occasion be very frustrating and confusing, for example when even a (I hope) healthy sceptic like me was tempted to fall into the memory-of-water trap when trying to crystallize ice III from the liquid. We may understand its essential structure, dynamics, and interactions in the crystalline, amorphous, liquid, and gaseous states. But it still presents many challenges for us to meet if we are to fully understand aspects of its functionality, especially in important physical, chemical, and biological systems.

Further reading

Perhaps oddly, good non-technical books and articles on water are few and far between—by far the best in terms of its wide coverage is Ball's *H_2O: A Biography of Water*. Eisenberg and Kauzmann's *The Structure and Properties of Water* may be over forty years old but still contains much good and illuminating material. The two series edited by Franks provide comprehensive coverage of most aspects of water, though again considerable advances in both data quality and understanding have been made since then.

Non-technical books and articles

P. Ball, *H_2O: A Biography of Water* (London, Weidenfeld and Nicholson, 1999).

J. L. Finney, 'Ice: the laboratory in your freezer', *Interdisciplinary Science Reviews* 29 (2004) 339.

J. L. Finney, 'Bernal and the structure of water', *Journal of Physics: Conference Series* 57 (2007) 40.

J. L. Finney, 'Sculpting ice molecule by molecule'. In H. Aardse and A. van Baalen (eds), *Findings on Ice* (Amsterdam, Pars Foundation and Lars Müller Publishers, 2007), 79.

J. L. Finney, 'The structure of water. A historical perspective', Journal of Chemical Physics 160 (2024) 060901.

F. Franks, *Polywater* (Cambridge, MIT Press, 1981).

More technical books

R. M. Daniel and J. L. Finney (eds), 'The molecular basis of life: is life possible without water?', *Philosophical Transactions of The Royal Society, London* B359 (2004) 1141.

D. Eisenberg and W. Kauzmann, *The Structure and Properties of Water* (Oxford, Clarendon Press, 1969).

F. Franks (ed.), *Water: A Comprehensive Treatise* (New York, Plenum, 1972 to 1982 (7 volumes)).

F. Franks (ed.), *Water Science Reviews* (Cambridge, Cambridge University Press, 1985 to 2009 (5 volumes)).

F. Franks, *Water: A Matrix of Life*, 2nd edition (Cambridge, The Royal Society of Chemistry, 2000).

R. M. Lynden-Bell, S. C. Morris, J. D. Barrow, J. L. Finney, and C. L. Harper, Jr (eds), *Water and Life: The Unique Properties of H_2O* (Boca Raton, CRC Press, 2010).

V. F. Petrenko and R. W. Whitworth, *Physics of Ice* (Oxford, Oxford University Press, 1999).

Reviews and scientific articles

P. Barnes, I. Cherry, J. L. Finney, and S. Petersen, 'Polywater and polypollutants', *Nature* 230 (1971) 31.

T. Bartels-Rausch, V. Bergeron, J. H. E. Cartwright, et al., 'Ice structures, patterns, and processes: a view across the icefields', *Reviews of Modern Physics* 84 (2012) 885.

J. D. Bernal and R. H. Fowler, 'A theory of water and ionic solution, with particular reference to hydrogen and hydroxyl ions', *Journal of Chemical Physics* 1 (1933) 515.

J. D. Bernal, 'The structure of liquids', *Proceedings of The Royal Society, London* A280 (1964) 299.

D. T. Bowron, J. L. Finney, A. Hallbrucker, et al., 'The local and intermediate range structures of the five amorphous ices', *Journal of Chemical Physics* 125 (2006) 194502.

P. G. Debenedetti and H. E. Stanley, 'Supercooled and glassy water', *Physics Today*, June 2003, 40.

J. R. Errington and P. G. Debendetti, 'Relationship between structural order and the anomalies of liquid water', *Nature* 409 (2001) 318.

J. L. Finney, 'Hydration processes in biological and macromolecular systems', *Faraday Discussions* 103 (1996) 1.

P. Gallo, K. Amann-Winkel, C. A. Angell et al., 'Water: A Tale of Two Liquids', Chemical Reviews 116, (2016) 7463.

P. Gallo, J. Bachler, L. E. Bove et al., 'Advances in the study of supercooled water', European Physical Journal E 44 (2021) 143.

T. Loerting, K. Winkel, M. Seidl, et al., 'How many amorphous ices are there?', *Physical Chemistry Chemical Physics* 13 (2011) 8783.

D. Marx, A. Chandra, and M. E. Tuckerman, 'Aqueous basic solutions: hydroxide solvation, structural diffusion, and comparison to the hydrated proton', *Chemical Reviews* 110 (2010) 2174.

M. J. Mottl, B. T. Glazer, R. I. Kaiser, and K. J. Meech, 'Water and astrobiology', *Chemie der Erde* 67 (2007) 253.

J. C. Palmer, P. H. Poole, F. Sciortino, and P. G. Debenedetti, 'Advances in Computational Studies of the Liquid-liquid Transition in Water and Water-Like Models', Chemical Reviews 118, (2018) 9129.

P. Gallo, K. Amann-Winkel, C. A. Angell et al., 'Water: A Tale of Two Liquids', Chemical Reviews 116, (2016) 7463.

C. G. Salzmann, P. G. Radaelli, B. Slater, and J. L. Finney, 'The polymorphism of ice: five unresolved questions', *Physical Chemistry Chemical Physics* 13 (2011) 18468.

C. G. Salzmann, 'Advances in the experimental exploration of water's phase diagram', Journal of Chemical Physics 150 (2019) 060901.

H. Savage, 'Water structure in crystalline solids: ices to proteins', *Water Science Reviews* 2 (1986) 67.

A. K. Soper, 'The radial distribution functions of water and ice from 220 to 673K and at pressures up to 400MPa', *Chemical Physics* 258 (2000) 121.

A. K. Soper, 'Structural transformations in amorphous ice and supercooled water and their relevance to the phase diagram of water', *Molecular Physics* 106 (2008) 2053.

A. K. Soper, 'Recent water myths', Pure and Applied Chemistry 82 (2010) 1855.

A. K. Soper, 'Supercooled water: continuous trends', *Nature Materials* 13 (2014) 671.

A. K. Soper, 'Is water one liquid or two?', Journal of Chemical Physics 150 (2019) 234503.

B. Stevens and S. Bony, 'Water in the atmosphere', *Physics Today* (June 2013), 29.

Index

Index